Routledge Re

Water Demand for Steam Electric Generation

In this book, first published in 1965, the authors identify the technological opportunities and costs of water recirculation and water quality adjustment in thermal plants, relating them to the possibilities for minimal expenditure and maximum efficiency in the use of water for servicing an entire region with thermal power. *Water Demand for Steam Electric Generation* will be of interest to students of environmental studies.

Water Demand for Steam Electric Generation

Paul H. Cootner and George O. G. Löf

First published in 1965
by Resources for the Future, Inc.

This edition first published in 2015 by Routledge
2 Park Square, Milton Park, Abingdon, Oxon, OX14 4RN
and by Routledge
711 Third Avenue, New York, NY 10017

Routledge is an imprint of the Taylor & Francis Group, an informa business

© 1965 Resources for the Future

The right of Paul H. Cootner and George O. G. Löf to be identified as authors
of this work has been asserted by them in accordance with sections 77 and 78 of
the Copyright, Designs and Patents Act 1988.

Publisher's Note
The publisher has gone to great lengths to ensure the quality of this reprint but
points out that some imperfections in the original copies may be apparent.

Disclaimer
The publisher has made every effort to trace copyright holders and welcomes
correspondence from those they have been unable to contact.

A Library of Congress record exists under LC control number: 65027669

ISBN 13: 978-1-138-85739-1 (hbk)
ISBN 13: 978-1-315-71873-6 (ebk)
ISBN 13: 978-1-138-85747-6 (pbk)

Water Demand for Steam Electric Generation:

AN ECONOMIC PROJECTION MODEL

by

Paul H. Cootner

and

George O. G. Löf

RESOURCES FOR THE FUTURE, INC.

1755 Massachusetts Avenue, N.W., Washington, D.C. 20036

Distributed by

THE JOHNS HOPKINS PRESS, Baltimore, Maryland

Resources for the Future is a non-profit corporation for research and education in the development, conservation, and use of natural resources. It was established in 1952 with the co-operation of the Ford Foundation and its activities since then have been financed by grants from that Foundation. Part of the work of Resources for the Future is carried out by its resident staff, part supported by grants to universities and other non-profit organizations. Unless otherwise stated, interpretations and conclusions in RFF publications are those of the authors; the organization takes responsibility for the selection of significant subjects for study, the competence of the researchers, and their freedom of inquiry.

This book is one of RFF's water resources studies, which are directed by Allen V. Kneese.

Director of RFF publications, Henry Jarrett; *editor*, Vera W. Dodds; *associate editor*, Nora E. Roots.

Preface

Managing the quantity and quality of the nation's water resources entails continuing and growing investment, much of it related to industrial development. By far the largest of the industrial demands on water is that of steam electric generation. Steam power plants withdraw over twice as much as all other industries combined, and most of this water is used for cooling. Recirculating the water decreases the amount needed but increases the cost of operations. At the same time, it can bring about internal economies that may considerably reduce the net cost of recirculation to the plant.

A further factor bearing upon the justification of recirculation is the disposal of waste heat. A plant can at little or no cost to itself discharge it directly into the source stream, thus increasing the temperature and lowering the quality of the water for other users. Or the plant may choose to build cooling towers which will transfer the heat into the air, where it will disperse and impose no costs upon successive users of the water course.

Those responsible for the management of water courses must balance the costs and gains associated with the provision of increased water supplies and the control of waste discharges—including waste heat. In order to do this rationally, they must have systematic information on the costs and gains that various water users experience if their supplies are augmented in quantity or quality or if their waste discharges are regulated. To help meet this need for information, Resources for the Future has initiated several studies of major water-using activities, of which this is one. Another, *The Economic Demand for Irrigated Acreage*, by Vernon W. Ruttan, was published this year by The Johns Hopkins Press. Other studies, dealing with the pulp and paper industry, the petroleum industry, and the beet sugar industry, are under way.

In this book the authors identify the technological opportunities and costs of water recirculation and water quality adjustment in thermal plants, relating them to the possibilities for minimal expenditure and maximum efficiency in the use of water for servicing an entire region with thermal power. The economic projection model they have devised takes into account the main factors that regulate the demand for water by steam electric plants and the relationship between these factors and the use of water. To illustrate the practical uses of the model, Mr. Cootner and Mr. Löf in a final chapter apply their methodology to a specific region—the West South Central comprising the four states of Arkansas, Louisiana, Oklahoma, and Texas—in order to estimate the region's expected water use for steam electric generating plants in 1980.

Those readers who wish to learn or review the technical principles of steam power production are referred to Appendix A, which presents an elementary exposition of the thermodynamics of electric power production. Similarly, those who wish to investigate the economic basis of the "supply curve for thermal efficiency" more thoroughly than the text provides will find a detailed explanation in Appendix B.

One of the authors is an economist, the other a chemical engineer. Paul H. Cootner is associate professor of economics in the School of Industrial Management, Massachusetts Institute of Technology; he received his B.S. in mathematics and M.A. in economics from the University of Florida, and his Ph.D. in economics from MIT. George O. G. Löf is a private consultant and also research associate with Resources for the Future. He holds a B.S. in chemical engineering from the University of Denver and a Sc.D. from the Massachusetts Institute of Technology.

ALLEN V. KNEESE
Director, Water Resources Program
August, 1965   *Resources for the Future, Inc.*

Acknowledgments

The authors wish to express their appreciation to the numerous people who offered suggestions and comments at various stages of this work. There were several who supplied engineering and economic information, including J. E. Downs, of the General Electric Company; M. A. Santalo, of the Massachusetts Institute of Technology; Stanley Moyer, of the Philadelphia Electric Company; and William Rose, Donald Cook, and John Tillinghast, of the American Electric Power Service Corporation.

Opinions and comments from the following were very helpful in revising the first draft of the manuscript: Charles D. Curran, of the U.S. Study Commission for Texas; Henry L. DeGraff and James J. Flannery, of the U.S. Public Health Service; Abraham Gerber, of the American Electric Power Service Corporation; R. A. Kampmeier, of the Tennessee Valley Authority; R. F. Kaser; Edward H. Miller, of the General Electric Company; Marc Nerlove, of the University of Minnesota; Vernon L. Smith, of Purdue University; and Lowdon Wingo, Jr., of Resources for the Future, Inc.

At a later stage of revision, comments and suggestions from J. B. Dickey, Jr., of The Marley Corporation; T. H. Hamilton, of the Fluor Products Company; and T. E. Larson, Brian Berg, and R. W. Lane, of the Illinois State Water Survey, were particularly useful. The authors appreciate the careful checking done by these individuals and acknowledge with thanks their contribution to this work. Responsibility for opinions and conclusions reached in this study is solely the authors', however, and complete agreement of all reviewers is not implied.

Special appreciation is expressed to Irving K. Fox, vice president of Resources for the Future, Inc., who headed the water resources program at RFF during the early part of this study, and to Allen V. Kneese, present head of the water resources group, who contributed the Preface and portions of the Introduction to this book. Suggestions and encouragement from these colleagues contributed in major degree to the performance and completion of the study. To them the authors express sincere thanks.

Paul H. Cootner

George O. G. Löf

Contents

Figures in Text

Figures in Appendix A

Figures in Appendix B

Chapter I

Introduction

The construction of facilities for managing the quantity and quality of the nation's water resources continues to command large and growing investment. Yet, those responsible for the planning and management of regional water resource systems are often inadequately equipped to estimate and forecast the benefits and costs associated with various existing and potential water uses. This is especially true of municipal and industrial water use and the use of water courses for waste disposal. That this situation exists is no reflection on the ability of water resource planners but rather reflects a comparative lack of research on these questions.

The result of this state of affairs has been that usually projections have been in terms of "requirements" derived from the application of current ratios or coefficients, possibly adjusted for technological change, applied to projected future output levels. These "requirements" do not reflect in any systematic way the influence of economic factors such as the cost of input and the costs associated with waste disposal. Since in many cases waste loads and inputs including water can be very sensitive to costs, it is important that methods be developed for forecasting these responses.

Use of water by industry has several aspects, which, from the point of view of over-all efficient use of the water resources of a region, can conveniently be divided into three categories.

First, industry *withdraws* huge amounts of water from water sources. In the United States as a whole, about twice as much water is withdrawn by manufacturing industries as by municipalities. Steam power plants, which are the focus of this study, withdraw over twice as much as all other industries combined. All except a negligible portion of the thermal withdrawals are for cooling. These huge withdrawals are possible at some locations only if stream flows are maintained by regulation. Also, the quality of the water available for withdrawal may significantly affect the costs of the withdrawing industry. On the other hand, it is possible in most instances for individual industrial clients to reduce withdrawals —by recirculation—and to improve the quality of intake water by treatment in the plant. Moreover, frequently lower water quality does not preclude use of the water resource but simply imposes additional costs on the using industry. Decisions to regulate the quality and quantity of water sources must therefore turn on the technological opportunities for, and the cost of, recirculation, on the opportunities for quality adjustment and on the costs of bearing poor quality in the water-using plant.

In the thermal power industry there are large-scale opportunities for recirculation—at a cost, of course. One of the major objectives of this study is to identify such opportunities and costs and to provide a model for projecting them into the future. Also, a means is devised for systematically taking into account the major type of costs imposed upon thermal plants by deterioration of the quality of the input water, i.e., reduced cooling efficiency because of temperature rise in the input stream.

A second dimension of water use by industry and a second type of impact on regional supplies occurs because of actual *depletion* or *loss* of water associated with use. In thermal power production, this is a variable, but within much narrower bounds than water withdrawal. Variation in losses is primarily a function of input costs other than water rather than costs associated with the use of water itself. The model devised in this study also permits losses to be derived and projected as a function of the costs of various inputs.

The final, and frequently most important, impact on regional water supplies results from *waste disposal* by industry. In the case of thermal power, the waste to be disposed of is heat. The waste heat can either be discharged directly into a surface water source or transferred into the air by means of cooling towers. Since the atmosphere contains a vast amount of dilution capacity, discharge to the air is harmless. The technology and economics of cooling towers are such that if they are used the water coming from them is recirculated to the plant, thus vastly reducing the water intake of the plant. Clearly then, there is a strong interdependency between opportunities for disposal of waste heat and water supply intake in this industry.

Because all but a negligible percentage of the water used in thermal electric plants is for cooling, the amount of water used for steam electric production depends both on the amount of cooling to be done (the cooling demand) and the technical efficiency with which the water is used in performing the necessary cooling (the withdrawal demand). These two aspects of water demand are largely determined by separate sets of factors. The cooling demand itself is determined by the central parameters of the process of electricity production. The withdrawal demand, on the other hand is determined, first, by the quantity and cost of water available; second, by the economics of heat exchange and recycling, and third, by any costs to the plant that may be associated with the disposal of waste heat. Since each of these elements of the problem is important in its own right, each is discussed briefly below, together with other key factors, as background for the more detailed discussion of the technology and economics of thermal generation presented in succeeding chapters.[1]

An example of how the methodology used in this study can be applied is given in Chapter VII, where a projection of water demand for steam electric generation for a hypothetical region is worked out, using procedures that can be followed by a planner in the field.

COOLING DEMAND

Cooling demand refers to the amount of heat, usually expressed in terms of British Thermal Units (Btu),[2] which must be dissipated by the cooling water. The heat which must be disposed of in this manner is discarded in the course of the generating process. Since the fuel itself is an expensive input in making electricity, and since disposing of the heat involves expenditure of additional funds, it is clear that there exists a strong incentive for electric utilities to conserve fuel (and therefore cooling demand for water). The fact that heat is still discarded despite so strong an incentive for conservation suggests that there are also substantial costs involved in increasing the efficiency of fuel use and that, after some point, these costs overbalance the economies of improved efficiency. As will be shown, there is also a theoretical technical requirement for discarding a portion of the heat.

---

[1] Readers who wish to study in greater detail the technical principles underlying steam electric power production are referred to Appendix A, which should be read before Chapter II. Similarly, those who are interested in the economist's techniques for deriving the supply curve for thermal efficiency are referred to Appendix B, which should be read after Chapter II.

[2] A measure of heat energy, equivalent to the amount of heat required to raise the temperature of one pound of water by one degree Fahrenheit.

The level of thermal efficiency chosen for new plants is customarily the result of a careful balancing of the costs and gains involved in the choice, although it turns out that at current prices the costs of water supply carry relatively little weight in making the calculation. The level of thermal efficiency chosen, together with the size of plant selected, is one of the two major elements in the investment decision concerning new generating plants, so we find that the demands of electric utilities for cooling water are intimately tied up with the whole investment process in the industry.

In order to forecast the regional demands for cooling, the nature of the electricity producing process was investigated in some detail. The production relationships involved in determining the level of thermal efficiency were studied, and the costs of supplying the various levels of efficiency were derived. In this study, these detailed relationships are summarized in a "supply curve of thermal efficiency" (Figures 1, 5, and 6) from which one can compare the costs of past advances in technology with price of future improvement. This device is of inestimable value in determining the likelihood of continued advance at the very rapid rates which have prevailed in the past.

These costs, as determined from the supply curve, are then compared with a similar curve reflecting the benefits in fuel and water savings which can be expected to accrue from such improvements. The latter curve, which we labelled "the demand curve for thermal efficiency" (Figure 9), intersects the previously established supply curve at the point of economically optimum levels of efficiency. Because, however, the level of fuel and water costs varies with the different conditions of fuel and water supply in the various regions of the country, there is a different "demand curve" for each set of such conditions. By comparing these various demand curves with the basic conditions of supply, we can determine the regional variation in the appropriate levels of new plant efficiency.

WITHDRAWAL DEMAND

Just as cooling demand is highly dependent on the productivity of the fuel that is used in producing electric power, withdrawal demand is primarily a function of the productivity of the water used to perform the cooling. Given any cooling load, the amount of water necessary to take care of it is a function of (1) the amount of heat carried away by each unit of water supply on a once-through basis, and (2) the number of times a given unit of water can be reused from the time it is first withdrawn from the aquifer to the time it is finally replaced. The amount of heat discarded per unit of water is a function of water flow rate through the condenser, which can be, at least theoretically, varied over a wide range. However, since flow control is an economically inefficient technique for conserving water in the utility industry, this variable tends to remain fairly inflexible. The reasons for this are treated cursorily in Chapter VI, but the primary influence on withdrawal demand is the economics of recycling.

Like the determination of cooling demand, the projection of the amount of recycling depends on an assessment of relative costs and gains. Water conservation involves capital and operating expenses. To be economic, these expenses must be balanced by concomitant benefits. The benefits must be calculated in terms of diminished water requirements and improved water quality in streams. There are a number of economic dimensions to lower water use: not only are the costs of withdrawal reduced, but there are lower raw water pumping costs and treatment expenditures, and the less quantifiable benefits accruing to production when interruptions due to unavailability are less likely to occur. The costs which tend to offset these benefits are the equipment and operating outlays that are a function of the desired temperature of cooling, the wet-bulb temperature of the air in the neighborhood of the plant,[3] the quality of the water, and the opportunity costs of power.

To deal with the problem of withdrawal water demand we use an approach not unlike that, summarized above, for estimating cooling demand. A schedule is constructed indicating the level at which recycling will become economic under various conditions of water cost, temperature conditions, and plant efficiency. From this information, local water engineers should be able to select the appropriate relationships and thus estimate the amount of recycling to be expected under the special conditions operating in their region.

EVAPORATION DEMAND, OR DISAPPEARANCE

A third factor of significance in the use of cooling water in steam electric plants is the evaporation or disappearance of water due to the heat it receives. This evaporation is closely related to cooling demand, because most of the heat transferred to cooling water (i.e., cooling

---

[3] The temperature to which a thin film of water can be cooled by contact with air of a particular temperature and humidity.

demand) is later rejected by the process of evaporation. This is the case whether once-through water use is practiced, with its return to lake or stream, or whether recycling with evaporative cooling is employed. Thus, evaporation demand is the factor of most significance in arriving at the true maximum capability of a water supply for cooling purposes.

TECHNOLOGICAL ADVANCES THAT COULD INFLUENCE FUTURE WATER REQUIREMENTS

The pattern of water use in the production of electricity from fuel could be substantially altered if any of the newer power generation techniques gain in general use. The gas turbine, which requires essentially no water, is already operating to generate power in relatively small units (5,000 kw). Similarly, diesel-powered generators of about the same size have been employed for some time, and these, too, practically eliminate the need for water cooling. At present, the costs of power from these systems are lower than from steam plants of the same size, but they are much higher than the costs of power generated by large steam plants. Heavy use of either system to meet large demands would, therefore, require technical and economic advances not now foreseen. It is considered more likely that while both systems, particularly the gas turbine, will be used extensively in small generating plants, they will comprise only a small fraction of the total installed capacity in the United States.

Magnetohydrodynamics involves the generation of electricity by the passage of combustion products, at extremely high temperature, through a magnetic field. Experimental use of this principle has been successful, and considerable effort is being devoted to its development. Ideally, a system developed from this principle could produce electricity at efficiencies higher than with the steam electric cycle, and without substantial quantities of cooling water. As another possibility, such a system could be combined with a steam plant: the fuel would first be used for power generation in the magnetohydrodynamic portions of the plant, from which the gases still at high temperature would pass to a conventional steam electric unit for additional power production. If some difficult problems can be overcome application of this principle may become important in the future, in which case water requirements for power production would be substantially reduced, but it is too soon to appraise the prospects for practical use of the system.

Widespread use of atomic energy in electric power generation will not materially alter the needs for cooling water because conventional steam turbines and generators are still needed. If anything, slightly more water per kilowatt-hour may be used in atomic plants because the high steam pressures and temperatures possible in plants employing the usual fuels may be considered unsafe. The lower steam cycle efficiencies resulting from these limitations would produce a greater condenser heat discard per unit of electricity generated, but it is believed that the difference will be comparatively small.

Another possibility for reducing future water requirements in steam electric power plants lies in the use of air-cooled rather than water-cooled condensers. Although they are not as efficient in summer, during most of the year air-cooled condensers could nearly match the performance of water-cooled units in many areas of the United States. The principal drawback to their substitution for water-cooled units is their much higher capital cost, due to poorer heat transfer coefficients and the larger surface areas required. The advantages of water-cooled condensers —in efficiency and over-all power generation cost—plus the savings in equipment investment, now dictate water cooling everywhere in the United States.[4] But if factors other than water control a plant's location, and consequently water should be unavailable for this purpose, or if the cost of the water should be extremely high, air cooling could be employed. The likelihood of such conditions becoming common in the United States, at least during this century, is too remote, however, to warrant serious consideration of the influences of air-cooled condensers on water demand.

[4] Air cooling is practiced to a limited extent in Great Britain (see Chapter VI, p. 56).

Chapter II

The Nature of Water Use
by Steam Electric Utilities

Several facts concerning water use in the electric utility industry are significant in the supply-demand relationships developed in this study. First, at present costs, water is not important enough as an input to affect production decisions or plant location more than marginally. This is partly because water is still relatively cheap and partly because water-saving devices can be used if water costs should rise severely. Second, although the usual measures of water quality, such as mineral and bacterial content, are not important in this use, temperature is an important element of the water's real cost to the user. Finally, water is a joint factor input with fuel, which is usually the major cost in producing steam electric power.

The amount of water used for boiler feed has little bearing on the total water consumed in steam electric generation, for since water as a source of steam must be highly purified, practically all of it is recovered in the condensing process and is reused. The industry's large water-using operation occurs in cooling and condensing the steam as it finally leaves the turbine after part of its heat has been extracted and turned into mechanical and electrical energy. This cooling process is a matter not of convenience but of economic necessity. As the steam is exhausted into the condenser and cooled by the water, it condenses to liquid, thus occupying a smaller space than it did formerly and producing a partial vacuum. This vacuum at the exhaust of the turbine permits the entering high-pressure steam to undergo greater expansion, and hence deliver more energy to the turbine rotor, than if it expanded only to the pressure of the atmosphere. Roughly, the over-all efficiency of producing useful work from high-pressure steam can be in the 40 to 50 per cent range if the exhaust steam is condensed by use of cooling water, whereas it will be only 30 to 40 per cent if the turbine exhaust is directly to the atmosphere (reference 3, p. 10–10).

The amount of thermal energy remaining in the steam at the exhaust point, together with the size and heat-transfer characteristics of the condenser, determine the amount of water that will be needed for cooling the condenser. This heat, which is then finally discharged from the plant into the water, is the residue of the heat input which is not turned into mechanical work in the turbine or otherwise lost. This discarded heat is the part of the total transferred to steam in the boiler which is not transformed by the turbine into useful work. In effect, it is "waste" heat, but it is waste in a technological and not an economic sense. At all times in the history of steam electric power generation it has been possible, from an engineering point of view, to make fuel use more thermally efficient than it has been. The decision not to utilize the fuel more completely results from weighing the potential savings in fuel costs against the increased capital costs that would be incurred. In the past, as the costs of fuel and capital have changed the efficiency of fuel use has changed with them, and in the process the use of water per unit of electrical output has declined concomitantly.

The intimate connection between fuel consumption and cooling demand can be illustrated in the following manner. Let us assume that a certain new steam power plant operates with a gross thermal efficiency of 40 per cent (reference 8, p. 17). This means that 40 per cent of the chemical energy in the fuel ("heating value" of the fuel) is converted to electrical energy output. The amount of heat energy equivalent to one kilowatt-hour of electric energy is 3,413 Btu. Since the assumed efficiency implies that each unit of energy input results in only 0.4 unit of energy output, 8,533 Btu of energy must be supplied in the fuel to produce the single kilowatt-hour of electric output. The remaining 5,120 Btu/kwh are not accounted for in the final product and are therefore dissipated as heat.

To trace these losses we must examine some of the elements of our 40 per cent efficiency figure. Sufficient fuel to liberate 8,533 Btu must first be burned in a furnace and the resulting heat largely transferred to the steam in a boiler. Although this heat transfer is accomplished at high efficiency, it is not a perfect conversion; about 10 per cent is lost in hot exhaust furnace gases and by other minor means (reference 54). Thus, 853 Btu are lost at this point.

The high-pressure, high-temperature steam leaving the boiler possesses potential energy which now may be converted into mechanical energy by causing the steam to expand against the blades of a turbine to turn a rotor. Here again, while the efficiency of conversion is high, there are inevitable losses. Some of the steam leaks around the blades, and some of the energy is lost through friction.[1] The resulting loss of useful work may account for about 10 per cent of the energy entering the turbine, or 768 Btu (reference 8, pp. 74–90). However, this energy remains largely in the steam which passes through the turbine and into the condenser without being usefully turned into mechanical power. This loss is due to what is technically termed irreversibility in the turbine, i.e., departure from a theoretically perfect engine.

There is another small loss in the conversion of mechanical energy from the turbine to the electrical energy delivered by the generator. Modern large generator efficiencies are in the 96 to 98½ per cent range (reference 8, p. 95), so 3,413/0.975 or about 3,500 Btu of mechanical work must be delivered to a typical generator per kilowatt-hour of useful output. Accordingly, 3,500 − 3,413 = 87 Btu of heat must be removed from the generator. In nearly all modern power plants, generators are cooled by a gas or liquid flowing in a closed circuit, from which this heat is removed by cooling water.

It is seen that the heat losses from the boiler furnace and the generator comprise only 940 Btu of the 5,120 Btu that must be wasted. The remaining loss arises from the fact that the internal energy in the steam leaving the turbine must be discarded simply because it is not in a form in which it can be utilized. To convert any more of it to work would require even more expansion than the steam undergoes in the turbine. Further expansion, however, would require even lower discharge pressure, achievable only by providing a lower temperature for condensation of the exhaust steam. Lower temperature is not economically possible if the coldest available water is already being used for cooling, and it is also physically impossible to achieve (by refrigeration) without the expenditure of more power than would be saved.

The lowest limits of temperature (and pressure) to which the steam can be brought are established by the lowest temperature heat "sink"

---

[1] Although tolerances between the blades and the casing are quite small, some leakage is unavoidable, especially in view of the high temperatures involved and the expansibility of the metals. The friction question is an interesting one, since the effect of friction is to create heat rather than lose it. But frictional effects in the steam expansion process simply result in some energy remaining in the steam rather than being converted to useful mechanical work.

available. If the exhaust steam is cooled and condensed by water, the lowest conceivable temperature for the system is that of the coldest water regularly available, plus a small differential to cause heat to flow from the condensing steam into the cooling water. It would also be conceivable to condense the steam with air, but under practical conditions on a year-round basis this operation would be at a higher temperature than water cooling, and it would require much larger and more expensive condensers. It would also be possible to exhaust the steam into a huge refrigerator, but the cost in terms of both dollars and power of running the refrigerator would far exceed the amount of the power saved. Moreover, the waste heat from the refrigerator must ultimately be discarded to water or the atmosphere, so the over-all energy transfer process via a refrigerator would have no technical advantage either. At the limit of the normal temperature of the earth, the large losses in the turbine exhaust of any steam electric plant are unavoidable. Heat discard in the condenser is necessary to the achievement of the maximum possible efficiency.

This does not mean, however, that steam plant efficiencies are immutably fixed at any given level. The lower limit of cooling water temperature effectively determines the exhaust pressure, but there is no similar limit on the upper temperature that can be achieved. At given exhaust conditions, it is possible to improve the *thermal* efficiency of a steam boiler-turbine-generator set by increasing boiler temperatures and pressures, although the economic barriers may be harder to overcome. What it does mean is that no matter what the boiler conditions are, it is technologically impossible to extract 100 per cent of the energy supplied to the steam by the fuel. All of the energy transferred to the steam that is left in it at, say, 100°F. exhaust temperature must be discarded,[2] even if the boiler and turbine themselves were perfectly efficient.[3]

[2] In order to convert *all* the energy received by the steam to mechanical energy in the turbine, i.e., have a 100 per cent heat-to-work efficiency, it would be necessary to expand the steam down to the condition of absolute zero temperature, −460°F., where there would be no energy left in it to discard.

[3] Even given the temperature limits within which the process must take place, there are many other obstacles to efficiency which narrow the horizons of possible technological change. Steam does not have the properties required for the most nearly efficient heat engine, although it has one property—economy—which more physically perfect fluids do not have. In addition, there are the more conventional obstacles to perfection. There is friction, leakage, heat transfer inefficiency, mechanical breakdown—the whole host of evils to which all machines fall heir. Together, these theoretical and practical obstacles assure us that cooling water will always be a necessary part of the steam-electric generating process.

In the case of our hypothetical steam plant (Table 1), we must discard in the condenser all of the heat from the fuel which is not otherwise transferred. Thus, 8,533 Btu from fuel minus 853 furnace and boiler losses minus 87 generator loss minus 3,413 converted to electric energy equals 4,180 Btu removed in the condenser by cooling water. Heat passes from the condensing steam through metal tubes into cold water circulated through the condenser. The amount of water that is needed depends upon the water temperature increase desired, which is in turn dependent on the size and heat transfer characteristics of the condenser. In a typical operation, a 10° water temperature rise might prevail (reference 3, p. 10–13) (i.e., 10 Btu would be removed by each pound of water, equivalent to 83.3 Btu per gallon). In that case some 50 gallons of water would be needed to perform the cooling task. A 15° water temperature rise would require the circulation of 33 gallons per kilowatt-hour generated. Since the 87-Btu generator loss also ultimately requires cooling water for heat discard, a 10° rise would require about 1 gallon additional water for this use; with a 15° rise, total water requirements would be about 34 gallons per kilowatt-hour for condenser and generator cooling.

In assessing the significance of these technical factors, it should be helpful to consider briefly the history and trend of steam power plant efficiencies in the United States. As recently as 1950, the average efficiency of central station steam power plants was about 24 per cent, corresponding to a fuel energy input of 14,200 Btu per kilowatt-hour generated. Even the very best performance of a large generating plant was 34 per cent, equivalent to a heat rate of 10,000 Btu per kilowatt-hour. These figures may be compared with the present central station *average* of 33 per cent (10,300 Btu per kilowatt-hour) and the present *best* performance of about 42 per cent (approximately 8,200 Btu per kilowatt-hour). (References 18 and 20.)

It is evident from the above figures that there have been remarkable advances in steam power technology in the past decade or two. Much higher steam pressures and temperatures, effective steam reheating cycles, larger installations permitting economical recovery of small energy losses, and a number of other improvements have permitted these efficiency increases.

However, further large increases in steam electric efficiency above the highest now realized are not likely because of the inherent thermodynamic limitations previously cited and because the benefits of still higher boiler pressures and temperatures must be of lesser magnitude. It has been predicted that by 1975 to 1980, the best power plant per-

TABLE 1. TYPICAL ENERGY BALANCE IN STEAM-ELECTRIC POWER PLANT

(Figures for 1 kwh net electrical output)

Assumed over-all efficiency	40%
Assumed generator efficiency	97½%
Heat equivalent of 1 kwh	3,413 Btu
Fuel energy required, 3,413/0.40	8,533 Btu
Heat losses from boiler furnace, at 10% of fuel use	853 Btu
Energy in steam delivered to turbine 8,533 − 853	7,680 Btu
Heat loss from electric generator, at 2½% of generator input	87 Btu
Electric generator output	3,413 Btu
Energy required for generator equals energy output from turbine	3,500 Btu
Energy remaining in steam leaving turbine, removed in condenser 7,680 − 3,500	4,180 Btu
Total cooling water required, 10° rise $\dfrac{4,180 + 87}{10 \times 8.33}$	51 gal.
Total cooling water required, 15° rise $\dfrac{4,180 + 87}{15 \times 8.33}$	34 gal.

formance should be an efficiency of about 45 or 46 per cent, corresponding to a heat rate of approximately 7,500 Btu per kilowatt-hour. This represents a further fuel energy savings, from present best usage, of only 700 Btu per kilowatt-hour, or about 8½ per cent improvement. Compared with the past changes in the best performance, from 10,000 in 1950 to 8,200 in 1962, further improvements in efficiency will raise the national average primarily by replacing old, inefficient plants with large, new plants whose performance may be only slightly better than the best of the present systems.

Similar conclusions may be drawn as to cooling water requirements. Assuming an arbitrary 10° temperature rise, the 1950 average water use in power plants would have been about 115 gallons per kilowatt-hour generated; the present average would approach about 60 gallons, and in the best plants, about 43 gallons. A future minimum requirement for approximately 40 gallons per kilowatt-hour can be foreseen. Corresponding figures for actual water loss through evaporation would range from an average of 1.1 gallon per kilowatt-hour in 1950 to present consumptions of 0.5 gallon average and 0.4 gallon minimum.

Two implications may be drawn from this technical review. First, the opportunities for decreasing the requirements for fuel and cooling water in the largest power plants are not as great as might be suggested by past improvements in performance. Sizable reductions in average fuel and water consumption are to be expected, however, through replacement of inefficient plants with large modern generation equipment.

Secondly, although thermodynamic principles limit the ultimate maximum heat-to-work efficiency to about 50 per cent with the steam cycle, there is some room for further improvements in present power generation practice. If economic considerations did not intervene, it would be possible to produce electricity with appreciably less fuel and water than now required. The economic obstacles to pushing very far beyond the present horizon are formidable, but it is nevertheless clear that more efficient techniques are well within our power to achieve. Indeed, looking back over time, the observer can note several occasions when the obstacles to further advance looked as difficult to surmount as they do today; yet under the pressure of economic motivation, they eventually yielded to engineering endeavor.

Chapter III

The Supply Curve
for Thermal Efficiency

The conclusion reached in the preceding chapter was that the optimum level of thermal efficiency to be designed into new steam power plants was determined not by technological considerations alone, but by a balancing of economic cost and advantage. Having established the existence of a range of flexibility in the choice of the desired efficiency, we now turn, in this chapter and the next, to an analysis of the important economic factors involved in that choice in light of the restraints imposed by technology. In the following pages, we shall investigate those economic and technologic parameters that affect the conditions under which increases in thermal efficiency are supplied. In Chapter IV we shall turn to the factors affecting the demand for thermal efficiency.

The phrase "the supply curve of thermal efficiency," as used in this study, specifies a functional relationship between the cost per kilowatt of capacity[1] of increasing the level of thermal efficiency[2] by one percentage

---

[1] In the practical, empirical task of determining a supply curve for thermal efficiency, there are almost always some decisions to be made about whether certain items should perhaps be deducted from revenues rather than added to costs or whether certain revenues should be considered to reduce costs or augment demand. Since the choice of procedure in these cases does not affect the final result, we are free to choose the most convenient approach. Because the development of a supply curve has proved to be the most difficult task in this study, the "costs" have been narrowly defined to be the pure equipment costs (of boiler, generator, turbine, condenser, etc.) of increasing thermal efficiency. All other changes in inputs are reflected in the demand curve. Thus, reductions in condenser costs associated with improved thermal efficiency are included as a negative item in the supply curve, but reductions in fuel costs because of the same improvement in efficiency are reflected only in the demand curve.

[2] For the purposes of this study, thermal efficiency will always imply the *percentage* that useful output bears to input, unless specified otherwise. Whenever the term "heat rate" is used, it refers to the number of Btu of fuel heating value required to produce one kilowatt-hour of electric energy.

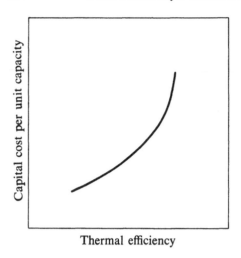

Figure 1. Supply curve (schematic) of thermal efficiency; unit capacity cost for marginal increases in thermal efficiency.

point and the level of thermal efficiency from which the increase is to be undertaken. The curve, shown graphically in Figure 1, implies that when efficiency is very low, it can be increased at a small additional cost, but as efficiency of plant is improved (i.e., as room for further improvement is reduced) the cost of making still further improvement increases. In the following pages we shall show that the supply curve does indeed follow this general shape (sloping upward to the right) and shall attempt to indicate its explicit form. Finally, we shall compare it to the *demand* for such efficiency improvements to determine the optimum level of improvement.

Details concerning the complicated process of developing the supply curve for thermal efficiency are to be found in Appendix B. The note below merely outlines the technique of derivation in order to provide some background for the subsequent discussion of the supply curve.

A NOTE ON HOW THE PRODUCTION FUNCTION WAS DERIVED. The first step in the procedure was a graphical portrayal of the engineering production relationships. Since the level of efficiency of a steam plant is an intricate function of a large number of engineering parameters, some simplification of the problem was necessary. The simplifying approach used was that commonly followed by industry engineers. The level of the less flexible parameters was taken as given, thus permitting the study to focus on the main factors affecting the level of efficiency, the temperature and pressure of the steam in the plant.

By use of temperature and pressure as the central factor inputs in the "production" of efficiency, the production relationships were portrayed in the form of iso-efficiency curves. For each level of efficiency, all the different combinations of temperature and pressure conditions which made that level of efficiency possible were connected by a smooth curve. Thus the production function was represented by a family of such curves, each representing a different level of efficiency (see Figure 2). These curves, calculated from engineering data and published estimates of industry engineers (references 3, 7, 8, 25, 28, 37), represent, to the best of our knowledge, the first formulation of the data in the form customarily used by economists, but the data themselves are the ones generally used in the industry. The curves should be considered as composites or rough averages of current plant performance, because there is plant-to-plant variation by virtue of such differences as types of fuel, cooling water temperatures, and the extent of heat economizing measures. For plants of more advanced design, especially those using double reheating of steam and boiler operation at super-critical conditions, additional curves must be added to the graphic portrayal, but the *principle* of the approach is not affected. Figure 3 illustrates these relationships for double reheating to 1,050°F (as explained in Appendix A).

It will be noted that the curves are all convex to the origin. This convexity is not an accident but reflects the "principle of diminishing marginal rate of substitution." Stated another way, the convexity of the curves is equivalent to the proposition that as more of one input (say temperature) is used, the quantity of the other factor which each successive unit of the first factor can replace diminishes.[3]

The material in Figures 2 and 3 merely specifies the state of engineering knowledge and technique. To develop the supply curve, which is basically an economic concept, we must have some additional information about the costs involved. To represent these economic data, we developed a set of curves similar to those representing the engineering relationships (references 17, 25, 51, 58). These "iso-outlay" curves are the locus of all the different combinations of steam conditions that can be supplied for any given sum of money. Unlike the iso-efficiency curves which are *convex* to the origin, these iso-outlay curves are *concave* (Figure 4). This implies that as temperatures are reduced and pressures are increased, suc-

---

[3] This "principle" is, of course, not an immutable "law," and it has its exceptions. At any rate, our functions obey the "principle" and this makes the manipulation of the relationships simpler than they might otherwise be.

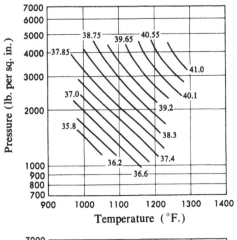

Figure 2. Iso-efficiency curves; one reheat to initial temperature.

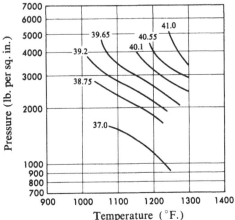

Figure 3. Iso-efficiency curves; double reheat to 1,050°F.

cessively larger decreases in temperature are needed to "buy" equal additional increases in pressure, and vice versa.[4]

With these two sets of curves it becomes possible to select the most efficient plant that can be constructed for any given sum.[5]

[4] This is equivalent to the proposition that the supply curves of temperature and pressure are upward-sloping, or to the more intuitive proposition that the more of an input that one buys, the greater the price that he must pay for it.

[5] The technical economic conditions for determining the most efficient plant for any given outlay, given the shapes of the curves, are as follows: the maximum level of efficiency is determined by finding the iso-efficiency curve which is just tangent to the given iso-outlay curve. The steam conditions for the plant are then determined by the temperature and pressure ordinates of the point of tangency. The theory behind this determination is discussed in reference 14 and, more elementarily, in reference 57.

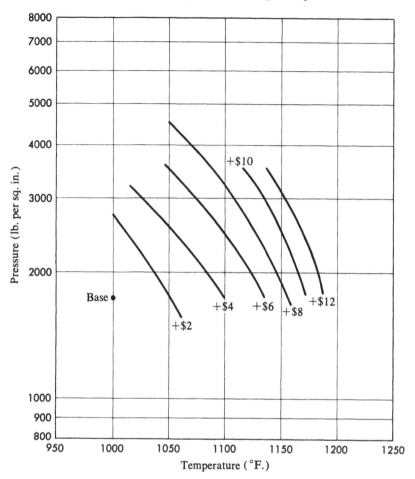

Figure 4. Marginal capital cost of increasing temperature and pressure; iso-outlay curves. (Measured from a base of 1,800 lb. per sq. in., 1,000°F.)

For each of several levels of investment per kilowatt of capacity, the maximum efficiency attainable may be read from the graphs (see Appendix B, Figure B-2). Replotting of these maximum efficiencies against the capital costs of achieving them shows that the cost rises at an increasing rate as efficiencies are raised. If the slope of this curve (i.e., the rate at which capital costs increase per unit rise in efficiency) is plotted against efficiency, we obtain the supply curve of thermal efficiency (Figure 5). This graph then shows the incremental capital outlays necessary to increase generating efficiency by 1 per cent, at any given level of efficiency.[6]

THE SHAPE OF THE SUPPLY CURVE

It will be noted in Figure 5 that although the supply curve which we derive by this technique is upward sloping throughout its length, the slope increases sharply in the neighborhood of 40 per cent efficiency. Since that level of efficiency is roughly that of the best new plants currently being installed, the sharp increase in costs in this area becomes important in determining prospects for improvement in efficiency.

The reason for this sharp bend in the supply curve lies in the metallurgical requirements for handling the temperatures and pressures associated with that level of thermal efficiency. In order to achieve those levels of efficiency economically, the temperature of the steam must rise to 1,200°F. or above, simultaneously with the maintenance of a

---

[6] In practice, the intermediate step has been eliminated by drawing iso-marginal outlay curves rather than iso-outlay curves.

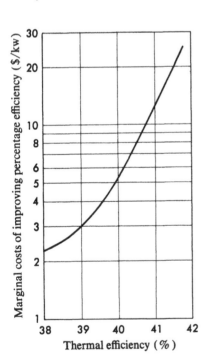

Figure 5. The supply curve of thermal efficiency in 1957.

high pressure. (See Figures 2 and 3.) While this temperature is not at the frontier of current equipment capability, it is comparatively high for industrial processes, and the combination of temperatures and pressures poses unusual structural problems. Even those problems, however, might not involve substantial cost increases if it were not for the long life required for turbine and boiler equipment.

Because ordinary alloy steels tend to deform ("creep") under high pressures at elevated temperatures, the metals that prove satisfactory for power generating equipment at temperatures below 1,200°F. are not adequate for the higher temperatures. To satisfy these new structural requirements, special high-temperature alloys are needed and these alloys require substantial additions of other expensive metals to the steel for the purposes of creep and corrosion resistance (references 3, 10, 16, 21, 64). As indicated in Table 2, these metals—nickel, chromium, molybdenum, manganese, etc.—are all substantially more expensive than steel. They differ widely in quality, and the prices given should be taken as illustrative rather than definitive.

Since the alloys now in use or contemplated may comprise 25 to 50 per cent of expensive non-ferrous metals, it can clearly be seen that those alloys will be much more costly than the materials used below the high-temperature barrier. This cost disadvantage is further amplified by the greater difficulty in making the alloys.[7]

[7] Since these alloys require exact proportioning of pure materials, they cannot be made by mass-production techniques, so they involve greater labor expense. In addition, their very strength makes them hard to fabricate.

TABLE 2. METAL PRICES

Alloying Element	Price, ($/lb.)
Nickel (electrolytic cathodes)	$0.79
Chromium (powder form)	1.15
Molybdenum (powder form)	3.15
Manganese (electrolytic)	0.2925
Carbon Steel	
Plates	0.0555
Tubes (base price)[1]	0.10
Stainless Steel	
Plates 304[2]	0.3681
Sheet 304[2]	0.40375

Source: American Metal Market, May 17, 1965.
[1] Base price. Actual boiler tube prices quoted on per foot basis and may be somewhat higher.
[2] Net of 5% discount in effect on May 17, 1965. Excludes extras.

THE EFFECTS OF TECHNOLOGICAL CHANGE

There is, of course, some possibility for reducing these costs in the future. As the alloys become more widely used, the costs of making and fabricating them will doubtless be reduced. The supply prospects for some of the alloying metals also foreshadow the possibility of some reduction in raw material costs. Nevertheless, the discrepancy between the costs of the two groups of metals is so great as to leave a residual increase which cannot be overcome. This is particularly true because fundamental micro-structure requirements place a lower limit on the amount of nickel in the alloy, which itself insures that alloys for use at high temperature and pressure will always be substantially more expensive.[8]

Based on an elimination of most of the extra compounding and fabricating costs and a reduction of 25 per cent in the costs of the alloying elements, we have estimated a second supply curve as we would expect conditions to prevail in the year 1980. While these conditions cannot be predicted with any great accuracy, they do have a basis in the expected future costs of production and the changed conditions of industrial organization, now coming to fruition, induced by recent expansion in the nickel industry. These considerations, investigated in more detail in Appendix B, suggest that the supply curve in 1980 will be as indicated by curve *B* in Figure 6.

Since technology has a way of changing in unforeseeable ways, it cannot be expected that these cost projections will prove to be faultless. In order that our supply curve for thermal efficiency will not be invalidated completely by deviations in one of the assumptions from future reality, we have added to Figure 6 two other curves indicating the costs of improving efficiency under alternative assumptions. If in the coming years scientific advances and diligent research serve to speed up the rate of cost reduction, a lower cost curve, such as curve *C*, can be utilized. It will be noted, however, that in all of the alternatives, the sharp increase in costs around current best levels remains.

A comment on the likelihood of technological change. If we distinguish, as economists commonly do, between changes in efficiency that take place because of knowledge of new ways of doing things, and changes that take place because the relative costs of different inputs change, we find that improvements in thermal efficiency, at least during the past

---

[8] The minimum percentage of nickel for austenitic steels is 18 per cent (reference 3, pp. 22-12, 22-14).

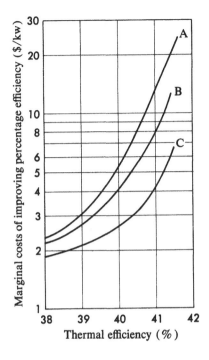

Curve A. This curve assumes no change in the supply of thermal efficiency and is identical with the estimated curve for 1957. This can mean either no change in current conditions or improved technology offset by higher prices.

Curve B. This curve, used in the projections made in Chapter VII, implies a 3 per cent annual improvement in real costs of supplying advanced plants and tapering off to a ¼ of 1 per cent rate of improvement at near current average efficiencies

Curve C. This optimistic curve assumes a 5 per cent annual real cost reduction in "frontier" plants and a 1 per cent annual reduction in more standard current plants. This assumes both an improvement in technology plus a secular decline in factor prices.

Figure 6. The supply curve of thermal efficiency in 1980. (1957 dollars.)

decade or so, have been due more to economic factors than to basically new technology. Today, there is the pure knowledge, if not necessarily the experience, for achieving substantial advances in thermal efficiency. Indeed, it is no reflection on the talents of industry engineers to point out that the theory and the metals with which to construct more efficient plants than those in existence have been available over most of the twentieth century. The real task has been to construct those plants at costs compatible with the profitable sale of power.

This is not to say that there has been no real technological change in the electric utility industry. To mention only a few advances, improved fabricating techniques have been found, knowledge of metallurgy has improved, and increased reliability of boilers has reduced the need for excess boiler capacity. Nevertheless, our central point remains that changes in the costs of fuel, metals, and labor, and increases in the quantity of power needed have been even more important in determining optimum thermal efficiency. This suggests that the element of unpredictability in technological change is probably less important in determining future costs of thermal efficiency supply than might be indicated by superficial study of the industry.

The gas turbine and atomic power. The fact that radical technological change has not been important in the past does not mean that it cannot be important in the future. At least two possibilities for major changes can be identified.

The gas turbine, like the steam turbine, is a device for transforming thermal energy into electricity. Unlike the steam turbine, however, it is powered by the combustion gases directly, instead of first transferring the combustion heat to steam. The primary advantage of the gas turbine in producing commercial electricity is its lower capital cost where only a few megawatts are required. It is therefore useful in very small power plants, or in large plants requiring operation only a small portion of the total time. It may become important in handling peak loads of utility systems.

Offsetting these advantages are two disadvantages, one mechanical and the other thermal. From a mechanical standpoint, the problem is turbine blade erosion. Since the powering force is the heated combustion products, this mixture is likely to contain any non-combustible impurities that were originally present in the fuel. While this is not an important problem if the fuel used is natural gas or, to a lesser extent, residual oil, the mineral impurities in coal have prevented the use of gas turbines in coal-fired stations. Oxidation and other high-temperature damage to blades and other metal components also may occur and adversely affect the life of the equipment.

The thermal problem arises from the fact that combustion gases, unlike steam, do not condense to a liquid when cooled. As a result of this, the pressure at the turbine exit must be above atmospheric pressure, and the exhaust temperature of the gas turbine is generally higher than it would be in the case of a steam turbine. Accordingly, for any given inlet temperature the thermal efficiency of the gas turbine will be lower. Offsetting this drawback, however, is the considerably higher heat input temperature possible in the gas turbine and the efficiency increase due to this factor. If the gas turbine and steam turbine are used in conjunction (the combustion gases give up part of their energy in the gas turbine before being used to generate steam for a conventional power plant), the combined efficiency is further improved because of a greater total useful temperature range in these two heat engines.

The implications of the potential adoption of the gas turbine are especially important for water use. Since there is no water cooling of the gas turbine exhaust, any electricity produced by the use of gas turbines will not involve water withdrawal or disappearance demand. In effect, the adoption of this innovation would substitute air cooling for water cooling in the electricity supply industry.

In view of its potential effects on the water input coefficient for electric power, is it likely that the gas turbine will significantly affect the utility production of the future? Based on the engineering and economic data that we have examined, the following developments seem to be most probable.

1) The major uses of the gas turbine for electricity supply will be in meeting small total demands, of a few megawatts, as in industrial plants and small communities where diesel-powered units may now be employed.

2) There is little likelihood that in the period covered by this study the gas turbine will be adapted for use by coal-fired stations.

3) Whatever gas turbine capacity is installed in large systems will be used to supply peaking power, and will as a result be operated at a very low load factor. That is, use of the gas turbine will be concentrated in that sector of electricity supply in which its capital cost advantage is most significant.

4) The gas turbine will be most intensively used in those regions where the capital costs of alternative peaking capacity is highest.

The potentialities for atomic power development pose less of a problem for water projections. There are two reasons for this. First, the "conventional" atomic power station, which utilizes a steam turbine to transform heat energy into electricity, introduces no radical changes into the nature of steam electric water use, and secondly, the projected costs of atomic power over the next two decades will prevent it from becoming a major factor in water use on economic considerations alone.

Despite substantial differences in design among the various existing and projected atomic power stations, they all utilize the same basic technique for the production of electricity. Heat from the atomic reactor is transferred to steam at high temperatures and pressures and the steam is, in turn, used to drive the turbine-generator. Conceptually, the sole differences between atomic and organic fuel stations are that the furnace is replaced by the atomic reactor and the transfer of heat to the steam usually takes place with an intermediate stage of heat exchange—hence the terms gas-cooled or water-cooled reactors.

Once the heat is transferred to the steam the two processes are the same. The steam expands through the turbine, and economic considerations require that it be condensed. Given a reactor and an organic fuel cycle of equivalent efficiency, the cooling demand of each will be identical. There are no major differences in the nature of water use.

But while atomic and organic fuel cycles of equal efficiency would have the same cooling demands, it is unlikely that atomic power plants will be operated at efficiencies as high as those prevailing in modern fuel-fired plants because of problems associated with high pressure and temperature inside a nuclear reactor. For an atomic cycle to be economically preferable to the optimum fuel cycle in a given location, it must offer lower fuel costs than the comparable cycle. This is because a nuclear reactor is more complex, and with its costly materials and radiation protection requirements, it is fundamentally more expensive than the conventional boiler and furnace, while the costs of the generators are identical and of the turbines not widely different. Thus, if the capital costs of the atomic plant must be greater than those for the conventional plant, it can be more economic only if its fuel costs are lower. In previous sections of this chapter, however, lower fuel costs have been shown to discourage increases in thermal efficiency; under these conditions, atomic plants would have lower efficiency and hence somewhat higher cooling demands (reference 19).

In terms of the scheme we outlined to analyze projected thermal efficiency, the increased capital costs of an atomic plant would be reflected in an upward shift of the supply curve for thermal efficiency. No change in the demand curve would be necessary, so long as the fuel cost could be computed in terms of cents per million Btu. In the present state of engineering knowledge about atomic reactors it is difficult to say precisely what the nature of the upward shift should be, but a few aspects can be detailed. Since the additional costs pertain principally to the boiler-furnace portion of the plant, they would not be perfectly proportional, and since the protective reactor shell is a significant part of the extra costs, there should be somewhat greater economies of scale in the atomic plants. In addition, because atomic plants have somewhat more heat exchange surface, they would be more sensitive to temperature considerations than conventional plants.

Overriding these surmises, however, is the view, held by most experts, that fission power will not become economically competitive with conventional energy sources during the next two decades in the United States except in some regions that have special fuel problems. In our projections, therefore, we have assumed that the reduced efficiency expected from the use of nuclear plants will be offset by the increased efficiency resulting from dual-cycle systems employing gas turbines with steam power units. Thus these two innovations are omitted from our forecasts for water use in the industry. Other methods for power generation, such as magnetohydrodynamic systems and various combinations of gas turbines and other non-condensing units, are considered to have negligible effect on total water demand during the next two decades. In some areas, expecially where natural gas is readily available and where power demands are low and scattered, the gas turbine might achieve considerable use. In these rather special situations, forecasts of water use for power generation would require adjustment, but on a regional or national basis there is little likelihood that they would be significantly affected.

Chapter IV

The Demand Curve
for Thermal Efficiency

In the preceding chapter, we set ourselves the task of developing a functional relationship between the level of thermal efficiency and the capital costs of achieving that level. To elaborate, our goal was a supply curve for thermal efficiency which could be set against the potential benefits from improvements in thermal efficiency in order that we might evaluate the prospects for further improvement. In this chapter, we turn to this second half of our task—the development of a schedule of *benefits* arising from changes in thermal efficiency. As was demonstrated in Chapter II, the inputs of water and fuel are so closely related that in assessing those benefits they must be evaluated together. But we are fortunate that these are the only important benefits of improving thermal efficiency[1] for, as a result, the demand curve proves to be less formidable to develop than the supply curve.

WATER AND FUEL COST PER KILOWATT-HOUR

The output of the electric generating process is measured in terms of kilowatt-hours. If the transformation of thermal into electrical energy could be accomplished with 100 per cent efficiency, only 3,413 Btu of heat energy would be required for the production of one kilowatt-hour of electrical output.[2] Since the measure of efficiency is the quotient of

[1] Actually, the benefits arising from the smaller condenser requirements could be viewed as an additional benefit, but because of the technological nature of the saving, we have preferred to treat those savings as a reduction in costs as indicated in the preceding chapter.

[2] The exact output required is 3,412.75 Btu, but for our purposes 3,413 will be sufficiently accurate. (See note 2, Chapter I.)

useful energy output and total energy input, thermal efficiency (expressed as a fraction) is

$$\frac{3,413}{\text{Energy input/kwh}} = H \qquad \text{(IV-1)}$$

where H is the over-all thermal efficiency of the process.[3] It then follows that energy input per kilowatt-hour, (B), is equal to

$$B = \frac{3,413}{H} \qquad \text{(IV-2)}$$

and that the fuel cost of a kilowatt-hour of generation (in cents) (K_f) is

$$K_f = \frac{0.003413f}{H} \qquad \text{(IV-3)}$$

where f is the cost (in cents) of 1,000,000 Btu of energy.

This formulation implies that fuel costs per kilowatt-hour will fall as either efficiency rises or price falls. In each case the change is proportional; a percentage change in one of the two independent variables implies the same percentage change in the dependent variable, although the signs of the changes will be different in each of the two cases. Graphically, the curves depicting the relationship between efficiency and fuel cost are a family of hyperbolas, one for each level of fuel price. Some of these curves are indicated in Figure 7.

In order to estimate the water costs associated with different levels of thermal efficiency, we must go a step further in analyzing the components of thermal efficiency. Because boiler feedwater represents so small a fraction of total water use in the thermal electric utility industry our analysis will concentrate solely on the use of water for cooling. Let us first review our explanation of cooling demand from Chapter II.

The real steam cycle is composed of two main parts: the boiler, where the heat is added, and the turbine, where it is turned into mechanical energy. Thus, in addition to the basic theoretical losses determined by the steam properties and operating conditions, there are further possible losses due to inefficiencies in both the heat input (boiler) or the conver-

---

[3] To read equations and applicable diagrams, see glossary of symbols at the end of this chapter.

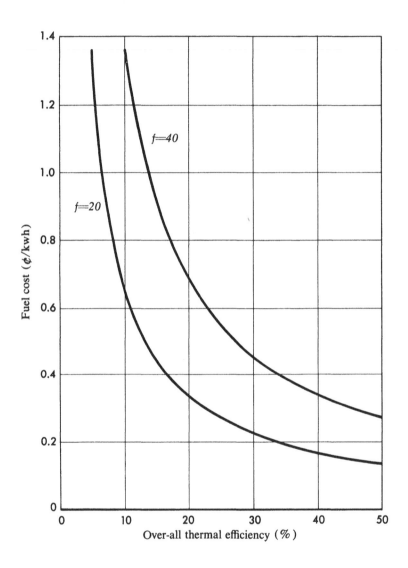

Figure 7. Fuel costs per kilowatt-hour (0.003413 f/H).

sion of the heat into the maximum theoretical mechanical energy.[4] Since an entering unit of energy is subject to all three possible efficiency losses, the total efficiency, H, is the product of α, the boiler efficiency, β, the turbine-generator efficiency, and η, the theoretical heat cycle efficiency.[5]

$$H = \alpha\beta\eta \qquad \text{(IV-4)}$$

For conceptual purposes, we can break the real steam cycle into two separate parts. In the first part the fraction $(1 - \alpha)$ of the energy in the fuel is lost to the outside air and $3{,}413/\beta\eta$ Btu are supplied to the turbine per kilowatt-hour generated. In the second part 3,413 Btu are transformed into one kilowatt-hour of output, so that the cooling load placed on the condenser and generator cooling unit is $h = 3{,}413/\beta\eta - 3{,}413 = 3{,}413\left[(1/\beta\eta) - 1\right]$. This then is the quantity of heat that must be removed from the power plant by the cooling water, per kilowatt-hour of plant electrical output.

In the process of condenser cooling (achieved by transferring heat from the hot substance, steam, to the lower temperature substance, cooling water), the steam condenses, changes to water, and gives up its latent heat of evaporation. Since it does not change state, the cooling water reflects the heat transfer merely by a rise in temperature. And recalling that a Btu is defined as the heat required to raise a pound of water by 1°F., the measure of the heat transferred to the cooling water is the simple product of its weight (in pounds) and its temperature change. Conversely, we can find the weight of water required for cooling purposes by dividing the heat load by the water temperature change designed into

[4] In the production of electric energy, there are further losses arising from the inefficiency of converting the mechanical energy of the turbine rotor into electrical energy in the generator. Although the amount of water use for steam condensation is a function of the losses before the generator, the generator efficiency is a factor in water usage *per unit of electrical output*. In large generators, efficiencies are in the 97 per cent to 99 per cent range (reference 8, p. 95). The 1 to 3 per cent loss is in the form of heat, which in large generating equipment is directly or indirectly transferred to cooling water. Thus, the losses from the generator join those from the steam cycle, all being dissipated in the plant cooling water. It is therefore convenient to consider the turbine and generator efficiencies together, their product, β, being the electrical energy produced by the generator divided by the theoretically recoverable energy (as work) in the steam supplied to the turbine.

[5] For many reasons (discussed in Appendix A), only a portion of this energy can be extracted from the steam in this cycle. This inefficiency is largely due to the properties of the steam itself and the design of the cycle rather than to mechanical inefficiency in the turbine or heat losses in the boiler. For the purposes of this chapter it is useful to distinguish these three elements.

the condenser and generator cooling unit. If we designate the temperature change by ΔT, the pounds of water required per kilowatt-hour are:

$$\frac{3{,}413\ (1 - \beta\eta)}{\beta\eta\Delta T} \ . \tag{IV-5}$$

A conventional unit for measuring water use is 1,000 gallons. Since a gallon of water weighs 8.33 lb., the amount of water (in thousands of gallons) required per kilowatt-hour can be obtained by dividing equation IV-5 by 8,330. If the price of water supplied to the plant is w cents per 1,000 gallons, the cost of cooling water will be [6]

$$K_w = \frac{0.41w(1 - \beta\eta)}{\beta\eta\Delta T} \text{ cents per kwh.} \ [7] \tag{IV-6}$$

WATER AND FUEL COST PER KILOWATT

So far we have dealt with fuel and water costs on a kilowatt-hour basis only. But thermal efficiency is "constructed" on a capacity basis: i.e., the costs of changing thermal efficiency are part of the costs of installing capacity, and these costs are not substantially altered by the degree of utilization of that capacity. In order to estimate the potential benefits of improving thermal efficiency, therefore, we must independently estimate the potential use of the capacity to be constructed. If a power plant were in operation throughout the year, it could produce a maximum of 8,760 kilowatt-hours per year for each kilowatt of generating capacity. Usually, however, the production from such a unit of capacity is something less than that, depending upon the *load factor* [8] for the system as a whole and the efficiency of plants already in the capital stock. If we take the plant load factor to be the fraction n, the total annual fuel cost (in dollars) will be equal to the cost per kilowatt-hour times the number of kilowatt-hours produced in a year by a kilowatt of capacity.

$$C_f = \frac{0.003413f}{H} \times 8{,}760n \times 0.01 = \frac{0.30nf}{H} \text{ \$/kw-yr.} \tag{IV-7}$$

---

[6] Water cost, w, may be fixed or variable. The unit cost of water withdrawn from a natural supply would usually not vary with quantity or temperature, but if supplied from a cooling tower lower operating costs for the tower during seasons when air temperature and humidities are low will decrease the value of w.

[7] This treatment assumes that generator heat is dissipated in the condenser along with heat from the condensing steam. Actually, separate cooling units are used, but generator losses are comparatively very small and water quantity requirements and water costs depend only on *total* heat removal by water.

[8] The load factor is the fraction of the total possible time that the plant is actually in operation.

Cooling water costs will be

$$C_w = \frac{0.41w \ (1 - \beta\eta)}{\beta\eta\Delta T} \times 8,760n \times 0.01$$

$$= \frac{36 \ wn \ (1 - \beta\eta)}{\beta\eta\Delta T} \ \$/\text{kw-yr}. \tag{IV-8}$$

Even these formulas, however, fall short of measuring the full fuel and water costs of a kilowatt of capacity. These are only the annual costs, and to facilitate the proper comparison with the capacity costs of supply, they must be summed over the life of the plant or the planning horizon of the firm, whichever is shorter. In computing these "lifetime" costs, we must take account of the fact that future costs are not the equivalent of present costs and of the fact that load factors do not remain constant over time. In view of the technological progress that historically has made new plants more efficient than old ones and the gradual increase of maintenance costs as plants age, the load factor can be expected to show a gradual secular decline, and in addition changing fuel costs will leave their cyclical imprint in that parameter. To make this annual stream of fuel and water costs properly comparable with the costs of building a unit of capacity of a given thermal efficiency, we must discount these outlays back to the time of initial installation, the discount factor, r, being the cost of money[9] to the firm at the time of installation.

In the most general case, the utility planning executives foresee, not a constant level of fuel costs or load factor, but rather a time function specifying those variables. One effect of introducing discounting into our curves representing fuel and water costs is to make the specification of the curves more difficult. It is now necessary to indicate not only the current load factor, but also its expected shape over time, and the same problem arises in the case of fuel. The decay curve for the load factor can be approximated by an exponential curve without doing violence to reality, and most projections of fuel costs are on a secular rate of change, rather than a cyclical basis. Partly as a result of this, but primarily because it simplified the analytic presentation, we will treat the load and fuel cost curves as if they were of the form[10]

---

[9] The evaluation of the appropriate "cost of money" is not an unambiguous task. For our purposes (cf. p. 45), the cost of money is the average weighted cost of debt and equity financing for the utility, prior to income taxes.

[10] The formulas may appear more forbidding to the non-technical reader than they are. To a close approximation they merely mean that fuel costs are expected to rise by a constant fraction x per year and that load factors are expected to fall by a constant fraction y per year. (If fuel costs are expected to fall, x would merely be considered a negative number.)

$$f = f_0 e^{xt}$$
$$n = n_0 e^{-yt}.$$

(IV-9)

Use of these relations permits us to investigate the effects of changes in the level of these factors separately from the complicating effects of changes in their time-shape.

This means that the fuel outlays in successive years will be

$$\frac{0.30 n_0 f_0 e^{x-y}}{H}, \quad \frac{0.30 n_0 f_0 e^{2(x-y)}}{H}, \quad \frac{0.30 n_0 f_0 e^{3(x-y)}}{H}, \quad \dots \text{etc.}$$

for τ years. Because of the cost of money, the present value of these amounts will be

$$\frac{0.30 n_0 f_0 e^{x-y-r}}{H}, \quad \frac{0.30 n_0 f_0 e^{2(x-y-r)}}{H}, \quad \frac{0.30 n_0 f_0 e^{3(x-y-r)}}{H} \quad \dots \text{etc.}$$

The sum of these τ payments is (from present value formulas)[11]

$$F = \frac{0.30 n_0 f_0 \left(1 - e^{-r'\tau}\right)}{H r'}$$

(IV-10)

where $r' = (r - x + y)$[12] and τ is the life of the plant. Thus, total fuel costs are still a hyperbolic function of the level of thermal efficiency and increase proportionately with increases in the initial level of fuel prices and plant factor. On the other hand, they decrease with increases in the adjusted rate of discount, though less than proportionately.[13] Expectations of rising prices increase the importance of the future in estimating fuel costs, while a secular decrease in expected level of operation reduces the importance of distant years.

[11] The presentation given above is heuristic, and uses a discrete approximation to a continuous process. Equation IV-10, however, is the exact formula, which could be derived directly by integrating: $\displaystyle\int_0^{\tau} \frac{0.30 n_0 f_0 e^{(x-y-r)t} dt}{H}$

[12] The practical significance of the new variable r' is to show what intuition may already tell us—that an expected secular rise in fuel prices tends to reduce the interest rate by the annual percentage inflation, while a declining rate of utilization raises the effective interest rate.

[13] If steam plants had infinite lives, the effect of changes in the discount rate would be exactly proportional. Given the rate of discount used by utilities and the length of plant life, it turns out that in fact the relationship is quite close to proportionality, which is another way of saying that $e^{-r'\tau}$ is fairly small.

In Chapter V we shall demonstrate that new steam power plants are almost always operated at close to 100 per cent load factor when new ($n_0 = 1$), so that differences in the average load factor over the life of a plant show up almost exclusively in a faster rate of decay in usage. The implication, which proves to be important in our study of the rate of introduction of new technology, is that investment planning of companies with high load factors is influenced as if they were operating with a lower rate of discount than others. While a similar case can be made for expected rates of change in fuel prices, fuel markets prove to be more interconnected (in space) than electricity demand is through time;[14] and thus there is less variation among utilities in this parameter.

In the case of water costs, a path of reasoning similar to that used in developing equation IV-10 results in a formula for capacity water costs of

$$W = \frac{36 w_0 n_0 \ (1 - \beta\eta) \ (1 - e^{-r''\tau})}{\beta\eta r'' \Delta T} \tag{IV-11}$$

where $r'' = (r - z + y)$, where z is the expected rate of increase in water prices.

THE DEMAND FOR THERMAL EFFICIENCY

The formulae that we have developed up to this point measure the cost of fuel and water expected to be consumed over the life of a unit of capacity. This item of total costs proves to be very important in the economic decision concerning the construction of capacity, but it is not the measure which is important in the determination of the level of thermal efficiency to be designed into the new plant. It is not the total level of such costs which determines the potential benefits from improving thermal efficiency, but rather the degree to which those costs can be reduced by such marginal increases. Economic rationality argues that so long as the cost decreases resulting from the reduction in heat requirements exceed the costs of the reductions, the process of improving efficiency should be pushed further. In Chapter III we developed estimates of the costs of marginal increases in efficiency. The appropriate

[14] Differences in the rate of change of fuel costs among localities are limited (though not prevented) by the possibility (and cost) of transporting fuel. Differing load factors among utilities are primarily the result of different time patterns of demand by consumers (summer vs. winter, day vs. night, etc.) and prove to be much more, though not completely, insensitive to price change.

measure of benefits for comparison is the rate of change of total fuel costs with respect to H, which turns out to be [15]

$$\frac{dF}{dH} = \frac{-0.30 n_0 f_0 \, (1 - e^{-r'\tau})}{H^2 \, r'} \, . \tag{IV-12}$$

The minus sign in front of the expression merely implies that the change in total fuel costs is negative, i.e., that those costs decline as efficiency increases. What is more interesting is that while a given percentage improvement in efficiency yields a *proportionate* reduction in *total* costs, it results in a more than proportionate decline in marginal fuel costs. Even in the case of total costs, our formulae implied that absolute increases of one percentage point in thermal efficiency resulted in successively smaller decreases in percentage (and absolute) fuel costs as efficiency improved; our new formula implies that the *marginal* value of improving thermal efficiency falls even faster. The elasticity of the demand curve[16] is constant, however; for each 1 per cent increase in thermal efficiency there is a 2 per cent decline in the price utilities are willing to pay for it. Figure 8 indicates the shape of the curve.

The marginal water cost benefits arising from improved thermal efficiency are similarly computed. As has already been indicated, those benefits are only indirectly a function of the over-all thermal efficiency. While total fuel costs are affected equally by a change in any of the three components of over-all efficiency, changes in the efficiency of the boiler have no effect on amount of water use per kilowatt-hour generated so long as the pure cycle efficiency and turbine-generator efficiency remain unchanged.

[15] Whenever we have an algebraic formula representing any quantity, we can give a similar formula for the rate of change of that quantity by the processes of differential calculus. This is what we have done in equation IV-12. The symbol, dF/dH is mathematical shorthand for the expressions, "the rate of change of F with respect to H," or "the change in F, (dF), corresponding, to a small change in H, (dH)." Since H is expressed as a fraction, the rate of change of F per unit change in H is thus the change in F for a 100 per cent change in H. For 1 per cent change in efficiency, the change in F becomes one one-hundredth the value from equation IV-12.

[16] The elasticity of demand with respect to price is an economic concept usually defined as the percentage change in quantity demanded due to (infinitesimal) change in the price. In this context we are borrowing the concept to indicate the percentage change in fuel savings given a percentage change in efficiency. Mathematically, the elasticity of a function, R, with respect to H is $E = (dR/dH) \, (H/R)$

In this case $R = \dfrac{-0.30 n_0 f_0 (1 - e^{-r'\tau})}{H^2 r'}$ and the elasticity (E) is -2 implying a 2 per cent decline in fuel *savings* for each 1 per cent increase in H.

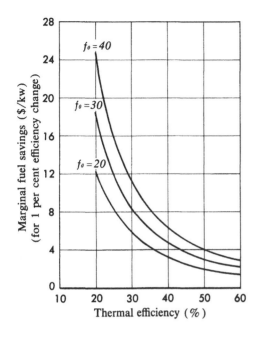

Figure 8. Marginal fuel savings in dollars per kilowatt-hour. ($x=0; y = 00.2; \tau = 35; n_0 = 1; r = 0.10.$)

However, since we expect almost all of future efficiency improvements to result from changes in η,[17] equation IV-11 may be differentiated to

$$\frac{dW}{dH} = \frac{-36 \, w_0 n_0 \alpha \, (1 - e^{-r''\tau})}{H^2 \, r'' \, \Delta T}. \qquad \text{(IV-13)}$$

In this event, the elasticity of the water benefits with respect to thermal efficiency is also 2, despite the fact that the total cost of water is a more complicated function of H than the total cost of fuel.

[17] Generally $d\beta/d\eta$ is negative if size of the turbine is held constant. This is because turbine losses are related to steam volumes and in high efficiency cycles given volumes have greater heat contents. The nature of $d\alpha/d\eta$ is more complex since boiler heat transfer is impeded absolutely by the metals associated with higher efficiencies but not necessarily more than proportionately to the higher heat content. Little accuracy is lost, in practice, from assuming that the cross-derivatives are zero.

For the more technically-inclined reader, equation IV-13 is derived as follows: If we multiply the numerator and denominator of IV-11 by α and set

$$C = \frac{36 w_0 n_0 (1 - e^{-r''\tau})}{r'' \Delta T}$$

we get $W = C \, (\alpha/H - 1)$. Differentiating, we get $dW/dH = C[-\alpha/H^2 + (1/H) \, (d\alpha/dH)]$. Since $d\alpha/dH = 0$, equation IV-13 follows immediately.

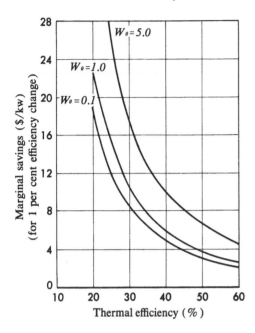

Figure 9. Demand curve for thermal efficiency; marginal savings, fuel and water, from improvements in efficiency. (f_0 = 30; x = 0; y = 0.02; z = 0; ΔT = 15; τ = 35; r = 0.10; n_0 = 1; α = 0.90.)

In the cases where the efficiency changes all take place in the pure cycle, the total demand curve becomes:

$$D = \frac{n_0}{H^2}\left[\frac{0.30 f_0\ (1 - e^{-r'\tau})}{r'} + \frac{36\alpha w_0\ (1 - e^{-r''\tau})}{r''\ \Delta T}\right] \qquad \text{(IV-14)}$$

The nature of the curve for some realistic values of the relevant parameters is shown in Figure 9.

GLOSSARY OF SYMBOLS

α = boiler efficiency (fraction)

β = turbine-generator efficiency (fraction)

B = energy input per kilowatt-hour (Btu)

C_f = yearly cost of fuel per kilowatt of capacity (dollars)

C_w = yearly cost of water per kilowatt of capacity (dollars)

D = reduction in cost from a marginal increase in efficiency (dollars per kilowatt of capacity)

e = 2.718

F = present value of total fuel costs per kilowatt of capacity (dollars)

f = fuel cost (cents per million Btu)

H = over-all thermal efficiency of electricity production (fraction)

η = theoretical steam cycle efficiency (fraction)

K_f = fuel cost per kilowatt-hour of generation (cents)

K_w = water cost per kilowatt-hour of generation (cents)

n = rate of utilization of capacity, i.e., load factor (fraction)

r = rate of discount of outlays

r' = adjusted rate of discount of fuel outlays

r'' = adjusted rate of discount of water outlays

τ = expected life of plant (years)

T = temperature (degrees Centigrade)

t = time (years)

W = present value of total water costs per kilowatt of capacity (dollars)

w = water cost per 1000 gallons (cents)

x = instantaneous rate of increase in fuel prices (fraction per year)

y = instantaneous rate of decrease in load factor (fraction per year)

z = instantaneous rate of increase in water prices (fraction per year)

subscript 0 = present time

Chapter V

Average Thermal Efficiency

The previous chapters have explored the factors determining the level of efficiency that would be selected by an economically rational design engineer. They are the motivations that are expected to affect the construction of plants at any particular point in time.

In looking ahead to the thermal efficiency of the electric utility system of 1980, however, we must take into account the longevity of previously constructed plant. Since the level of thermal efficiency of new installations is substantially above that of the stock of generating capacity, it is clear that even with no further improvement in best heat rates over the next two decades, it would be possible to increase efficiency by over 25 per cent. This situation is true today, but it has also been true at every point of time in the recorded history of the electricity industry. Figure 10 which shows the average thermal efficiency of U.S. utilities and the efficiency of the best plants equipped by a leading turbine manufacturer,[1] illustrates the gap between best practice and historical practice (reference 20).

In addition, the graph shows another interesting facet. While the curve representing best practice sometimes departs from trend, the relative stability of the upward movement is most marked. The curve of average efficiency, however, is much more variable—one might even say cyclical. A projection made during the World War I years would certainly have underestimated the future gain in efficiency, just as estimates made in the closing years of the expansion of the twenties

---

[1] These are the *plant* efficiencies which are estimated (by the manufacturer) to correspond to the efficiencies of *turbines* shipped in the year indicated. While these figures cover a large proportion of total industry shipments and are representative of that total, there is some indication that leadership in the construction of most efficient plants has alternated between the two leading manufacturers from time to time.

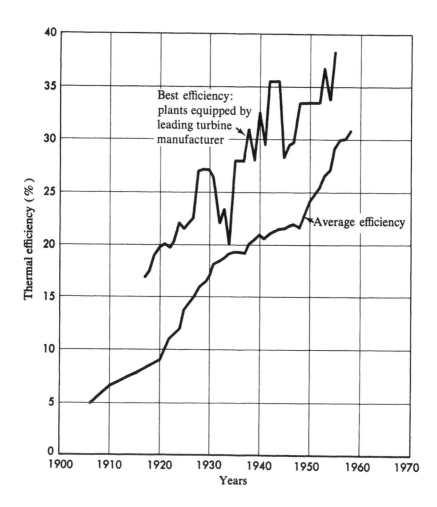

Figure 10. Average efficiency of U.S. utilities and of the best plants equipped by a leading turbine manufacturer.

would most likely have overestimated the potential improvement in succeeding years. A projection of the 1907–20 rate of growth would have been 33 per cent too small by 1940. A projection at the 1920–31 rate of improvement would have been 60 per cent too high by 1951. Even a projection with the entire 1907–45 period as a base would have already proved 30 per cent too low. This greater variability of the efficiency of the stock can only result from changes in the rate at which the new efficient capacity is added to existing capacity. This properly suggests that the rate of growth of demand and capacity in a region is a vital parameter in the estimation of future levels of efficiency in that region.

Since all available estimates of the growth of demand for electric power indicate a very rapid rate of growth over the next two decades, the efficiency of the stock of generating capacity in 1980 will be much higher than would be the case in a more stable industry. Still, even if we make the optimistic estimate that demand for electric power[2] (and capacity to produce that power) doubles in each of the next two decades, 15 per cent of the 1985 capacity would be already either built or under construction. On the other hand, if we were to expect 1985 capacity to be only three times the current level, the proportion of existing capacity still in use would be 20 per cent of the total (reference 18).[3] Since older plants are operated at lower plant factors than new ones, these proportions of capacity are undoubtedly larger than the corresponding proportions of production. Nevertheless, the importance of the existing stock in two-decade projections of long-lived generating capacity is clear.

In addition to changing rates of demand growth, however, there are *other factors* involved in the non-conformity of best and average efficiency. A study of the demand for electric power does not reveal fluctuations in demand of the kind that would account for the waves in the construction of plant capacity that have occurred. Those irregularities in the course of generating investment which *have* occurred largely arise from variations in the amounts of unused capacity held by utility plants. The nature of those fluctuations is shown in Figures 11 and 12, both before and after the removal of the basic trend. The underlying downward trend results from the growing density of electricity production, the increasing capacity and decreasing cost of energy transmission, and the superior reliability of modern machines. All of these factors either reduce the absolute need for emergency supplies or improve the capability for, and the economy of, drawing on neighboring utilities.

---

[2] In this chapter, we use the term "demand" to mean peak, one-hour demand for electric power. Because electric utilities must stand ready at all times to supply electricity as the consumer wants it, it is this peak demand which largely determines the required generating capacity.

[3] These figures assume that only plants built within the last ten years will still be in existence in 1980. This is about in line with historical experience, but is probably quite conservative in view of the slower expected obsolescence of plant in the future. It is obsolescence rather than physical deterioration which tends to govern plant retirement.

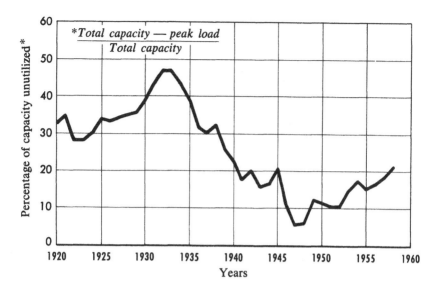

Figure 11. Unutilized capacity of electric utility industry.

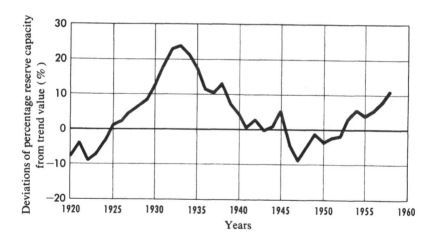

Figure 12. Unutilized capacity of electric utility industry; deviations from trend. (Trend fitted to 1920–30 and 1946–55. The trend is due to increased density of demand and improved transmission. Since demand growth and transmission line construction were both retarded in the thirties and during World War II, the deviations in those years are not directly relevant.)

As a result of these fluctuations in unused capacity, there have in the past been *substantial divergences between the growth of output and that of demand*. From 1929, when the level of utilization of capacity was at its post-World-War-I low of less than 65 per cent (ignoring the involuntarily lower levels in the early thirties), it was possible for the industry to expand capacity at 2¼ per cent less per year than peak demand over a period of eighteen years, to 1947 (reference 18). Since demand was growing at almost 5.5 per cent per year in this period, capacity was growing at a rate only 60 per cent of that of demand, and *this difference accounts for most of the retardation in the improvement of efficiency that can be observed* in Figure 10.[4] Similarly, the postwar rebound from the high utilization ratios of 1947 (94 per cent) to the 1958 levels of 78 per cent resulted in a growth rate of capacity 20 per cent greater (on an annual basis) than was necessary merely to meet demand, and added impetus to the postwar improvement in efficiency.

Thus, when capacity is growing faster than demand, the increase in average efficiency that results tends to be above the maintainable rate of improvement; that is, it tends to increase faster than would be expected if the same rate of technological and economic change were to persist.

Nevertheless, if a new wave of retrenchment were to reduce margins once more, they could not fall as far in the future as they have in the past because of the lower levels from which the decline would start. Furthermore, it is not likely that they can be reduced below the levels reached in the postwar years, particularly because the rapid development of air conditioning in recent years has tended to reduce the free time in the summer that was previously used for repair and maintenance. This means that more extra capacity must be allowed for the possibility that some maintenance will have to be undertaken in periods of high demand.[5]

[4] There was, in addition, a minor decline in the rate of improvement of the best new plants.

[5] Additional difficulty is caused by the nature of the measure of national capacity utilization. Since the very large majority of utilities have their peaks in December (until recently all of them did), the national peak is usually taken to be the sum of the peaks of the individual companies in December. At the present time, however, many utilities have their peaks in the summer, so that listing only the highest December demands understates the demands on their systems. Thus, the size of the December reserves overstates the excess capacity.

Partially offsetting this is the fact that the individual system peaks do not occur at precisely the same time, so that the sum of the December peaks is somewhat greater than the maximum demand at any one period in December. These differences between the timing of the peak loads of adjacent utilities results from time zone, geographical and meteorological differences.

What is the implication of this for the future? We can see it more clearly by looking at the experience of the post-World-War-II years. We have noted that some of the improvement in thermal efficiency in the postwar years was due to the renovation of the stock of generating capacity. If the rate of change of "best" thermal efficiency remains constant, it will be impossible to maintain the postwar rate of improvement in average efficiency unless greater and greater amounts of excess capacity are constructed. If, as we have indicated in Chapter III, the rate of change of new plant efficiency declines or ceases, it will be necessary for increases in excess capacity to be even greater in the coming decades if the postwar rates of improvement in average efficiency are to be maintained. On the other hand, if the sharp increases in reserves since the war prove to be only temporary, and utilization rates rise to their long-run trend level, the rate of improvement in fuel use will be much less than we have recently experienced. Furthermore, if this is true for the whole nation, it will have to be true to a greater extent in some *parts* of the nation, since the average presumes that some regions have greater reserves and some regions smaller ones. A region enjoying greater margins than the current national average would possess enough flexibility in meeting future loads to endanger the accuracy of any forecast that failed to take account of such flexibility.

Will reserves continue to increase; will they maintain their present level; or will they return to the prewar trend? In forecasting the trend of average efficiency, this matter appeared to be so decisive a variable that a thorough examination of the factors affecting reserve capacity requirements was undertaken as part of the background research for this study. In the course of that investigation it was found that the proportion of capacity in excess of peak demand (corrected for trend) was highly correlated with the cost of money to public utilities, the cost of money being defined as the weighted cost of an integrated financing (bonds *and* common and preferred stock) at current rates in the money market.[6] This was in full accord with our hypothesis concerning the investment problem: that utility decisions to carry excess reserves are made on rational grounds, balancing the advantages of carrying such capacity inventories and the capital costs involved. The economic advantages are the opportunity to schedule maintenance most efficiently, the ability to secure new industrial customers against the

---

[6] This is approximately equal to the "required rate of return" used by utilities commissions in setting rates. Because of the differential tax treatment of the several elements of costs, the returns are compared on a pre-tax basis.

competition of other utilities, the avoidance of lost revenues (and bad publicity) from service interruptions, the economies of scale involved in the construction of larger new plants, and the reduction in costs arising from the replacement (without demolition) of generation from inefficient plants. The offsetting disadvantages are the capital costs of maintaining the excess capacity and the increased risk to the enterprise from obsolescence or financial difficulty. Since the capital cost variable is the one most subject to change, variations in the level of such costs should prove to be the most important determinant of the reserve capacity.

That this theory seems to be substantiated by the facts about disposal suggests that past fluctuations have been regular and predictable variations around a generally declining trend in required excess capacity. (The declining trend seems, in turn, due to a reduction in the economies of scale and a reduction in the costs and the increased availability of power from other utilities in emergencies.) In particular, most of the recent postwar increase in the percentage of reserve margins is shown to be a temporary phenomenon which will not continue unless the current trend toward lower equity costs is maintained and/or the secular rise in interest rates reversed. While it would be foolhardy to attempt to predict a precise level for money costs in the future, there is no reason to assume that the level of reserves that will be considered desirable in 1980 will differ substantially from the trend level. This level, which we assume will be about 10 per cent by 1980, implies that capacity will rise at a level substantially below that of peak demand in the next two decades, as current levels of reserves are reduced.

COMPUTING AVERAGE EFFICIENCY IN 1980

Once we specify the expected course of reserve capacity over the next two decades, a few straightforward assumptions permit us to estimate average thermal efficiency in 1980. First, we assume that growth of capacity and improvement in thermal efficiency will take place at a steady geometric rate. As far as the latter assumption is concerned, it is validated by both historical data (see Figure 10) and the nature of our projections of efficiency. In the past, small incremental improvements in knowledge and in ability to turn that knowledge into usable product have, in fact, taken place, and there is little reason for this to change in the future since the horizon holds no important technological changes that would radically affect the economics of thermal efficiency. The assumption of a steady growth in capacity must be made with somewhat less assurance, however, for both our investigation and an independent study conducted by the National Bureau for Economic Research suggest that improvement has taken place in rather long cycles (reference 62). Nevertheless, the amplitude of fluctuations in this industry has been substantially less than that we are accustomed to note in the manufacturing industries, and since we have no reliable way of forecasting the cycles if they will, in fact, persist, a prospect of steady growth seems at least reasonable. Similarly, while our final projection of demand growth, like that of most industry students, forecasts a slower rate of capacity growth than has been experienced in the postwar period, it is impossible to predict precisely when the decline in the rate of growth will start; hence, we postulate a steady rate starting with the present.

The only other assumption that must be made concerns the general shape of the industry "load curve." While the peak demands upon electric generating capacity persist for only a brief time span, perhaps an hour or even less, certain basic loads persist 24 hours a day, 8,760 hours a year and intermediate demands last for a varying number of hours during the year. This information for a "typical" electric utility is summarized in Figure 13. On the vertical scale is measured "system load," that is, the amount of capacity being utilized at any instant of time. The horizontal scale is in hours. Starting from the left, we first plot, on the vertical scale the maximum amount of capacity, $\bar{G}$, used by the system for any period in the year. Corresponding to any other load less than $\bar{G}$, say G_1, we would plot a point indicating the number of hours in which the system bore a load equal to or greater than G_1. Finally there is some basic load, D, which is in operation throughout the entire year. It should be noted, furthermore, that the curve can be interpreted another way—as a graph of the precise amount of generation taking place in the hour indicated on the horizontal scale, where the hours have been rearranged in order of size of load. Looked at from this latter point of view, we can see that the area underneath this curve represents the total amount of electricity generated per year. If the "load curve" were a horizontal line it would mean that the utility would produce steadily at full capacity (except for its reserves) all year round. Since the curve is not horizontal a typical utility operates at much less than full theoretical production although, unlike its reserve capacity, this is not a matter of choice.

Faced with this overcapacity due to the nature of customer demand, utilities try to meet the demands with the greatest possible efficiency. Since the plants used to satisfy the base demands will operate at close to

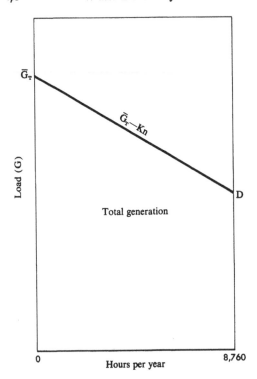

Figure 13. Load curve of a typical electric utility industry.

full capacity, it is economically wisest to use the most efficient units to supply these particular loads. For each load that is greater, the next most efficient plant is chosen, and for the peak loads, the least efficient plants are used. Thus, the average efficiency of production depends not only on the efficiency with which plants are constructed, but on the rate at which they are utilized. That rate in turn depends upon the load curve.

Historically, the tendency of the plant utilization rate has been to rise. This has partly been due to the price encouragement that utilities have given to off-peak use, and partly to such technological changes as the growth of air conditioning. But while this results in a load curve tending to occupy more of the theoretical "square" that spells full utilization, the curve itself has continued to be of the general shape indicated. In the model that follows we shall assume that this shape will continue into the future.

Given these assumptions, none of which is far removed from reality, the average efficiency in year T will be:[7]

[7] See the appendix to this chapter for derivation.

$$\overline{H}_T = \frac{H_T}{(a+1)\,L_T}\left[1 - \frac{(S_T - 2L_T + 1)^{a+2} - (1 - S_T)^{a+2}}{2(S_T - L_T)(a + 2)}\right]$$

$$(V\text{-}1)$$

where

$a = \dfrac{b}{d} = \dfrac{\text{average annual rate of growth of best efficiency}}{\text{average annual rate of growth of capacity}}$

$S_T = \overline{G}_T/C_T$ = ratio of peak load to total capacity in year T

$\overline{G}_T$ = peak load in the year T

C_T = generating capacity in the year T

L_T = load factor in year T.

In Chapter VII we develop our own estimates of these parameters for use in a sample set of projections for a specific region. Development of the above formula will permit other workers in the field to adjust the projections for different assumptions that may apply to the particular region in which they are interested.

Appendix to Chapter V

DERIVATION OF EQUATION V-1

To compute the average efficiency, $\bar{H}_T$ of a power system at time T, we must know the efficiency, $H(y)$, of each unit of capacity, y, and the number of hours that unit is operated, $n(y)$. Mathematically, it is

$$\bar{H}_T = \frac{\int H(y) \cdot n(y)\, dy}{\int n(y)\, dy}. \tag{V-1A}$$

To arrive at an expression for $\bar{H}_T$ we must first identify y more precisely. Take it to be yth most recent unit of capacity installed; i.e., the yth unit of capacity is older than $y - 1$ units. Let us further assume that capacity, C_t, increases at an exponential rate (d) so that

$$C_t = C_0 e^{dt} \tag{V-2A}$$

Then the yth unit of capacity at time, T, was installed in a year, m, identified by the following expression:

$$y = C_T - C_m = C_0(e^{dT} - e^{dm}) \tag{V-3A}$$

$$e^{dm} = \frac{C_T - y}{C_0} \tag{V-4A}$$

$$m = \frac{1}{d}\ \ln \frac{C_T - y}{C_0}. \tag{V-5A}$$

If we further assume that best efficiency grows at an exponential rate, b, then

$$H_t = H_0 e^{bt} \tag{V-6A}$$

and:

$$H(y) = H_0 e^{\,b\left(\frac{1}{d}\,\ln\frac{C_T - y}{C_0}\right)}$$

$$= H_0\left(\frac{C_T - y}{C_0}\right)^{b/d}. \tag{V-7A}$$

Referring back to the load curve (Figure 13), as long as y is less than $(\overline{G}_T - 8{,}760K)$, that unit of capacity will be operated a full 8,760 hours. If y is greater than $(\overline{G}_T - 8{,}760K)$, the relationship between the load and the hours of operation is

$$y = \overline{G}_T - Kn \qquad (\text{V-8A})$$

$$\text{or} \quad n(y) = \frac{\overline{G}_T - y}{K}. \qquad (\text{V-9A})$$

We can now try to evaluate equation V-1A. The denominator is obviously equal to total production, P_T, and is equal to

$$8{,}760\overline{G}_T - \frac{(8{,}760)^2}{2} K = 8{,}760 L_T C_T \qquad (\text{V-10A})$$

where L = the load factor for the system ($L_t = P_t/8{,}760C_t$). Note that

$$8{,}760K = 2(\overline{G}_T - L_T C_T) \qquad (\text{V-11A})$$

$$\overline{G}_T - 8{,}760K = 2L_T C_T - \overline{G}_T. \qquad (\text{V-12A})$$

The numerator of (V-1A) can now be written

$$\int_0^{\overline{G}_T - 8{,}760K} 8{,}760 H_0 \left(\frac{C_T - y}{C_0}\right)^a dy + \int_{\overline{G}_T - 8{,}760K}^{\overline{G}} H_0 \left(\frac{C_T - y}{C_0}\right)^a \frac{(\overline{G}_T - y) dy}{K}$$

$$(\text{V-13A})$$

$$= \frac{8{,}760 H_0}{C_0{}^a} \left[\int_0^{2L_T C_T - \overline{G}_T} (C_T - y)^a\, dy \right.$$

$$\left. + \int_{2L_T C_T - \overline{G}_T}^{\overline{G}} \frac{(C_T - y)^a (\overline{G}_T - y)\, dy}{8{,}760K} \right]. \qquad (\text{V-14A})$$

where $a = b/d$.

The first integral is equal to:

$$\int_0^{2L_T C_T - \overline{G}_T} (C_T - y)^a dy = \left[-\frac{(C_T - y)^{a+1}}{a+1} \right]_0^{2L_T C_T - \overline{G}_T}$$

$$(\text{V-15A})$$

$$= \frac{-[\overline{G}_T - C_T(2L_T - 1)]^{a+1} + C_T{}^{a+1}}{a+1}.$$

The second integral is equal to:

$$\frac{1}{8,760K} \int_{2L_TC_T-\overline{G}_T}^{\overline{G}_T} (C_T - y)^a (\overline{G}_T - y) \, dy$$

$$= \left[\frac{-(\overline{G}_T - y)(C_T - y)^{a+1}}{8,760K(a+1)} \right]_{2L_TC_T-\overline{G}_T}^{\overline{G}_T}$$

$$- \frac{1}{8,760K} \int_{2L_TC_T-\overline{G}_T}^{\overline{G}_T} \frac{(C_T - y)^{a+1}}{a+1} \, dy$$

$$= \frac{0 + 2(\overline{G}_T - L_TC_T)[\overline{G}_T - C_T(2L_T - 1)]^{a+1}}{8,760K(a+1)}$$

$$+ \left[\frac{(C_T - y)^{a+2}}{8,760K(a+1)(a+2)} \right]_{2L_TC_T-\overline{G}_T}^{\overline{G}_T}$$

$$= \frac{2(\overline{G}_T - L_TC_T)[\overline{G}_T - C_T(2L_T - 1)]^{a+1}}{8,760K(a+1)}$$

$$+ \frac{(C_T - \overline{G}_T)^{a+2} - [\overline{G}_T - C_T(2L_T - 1)]^{a+2}}{8,760K(a+1)(a+2)}. \qquad \text{(V-16A)}$$

But $8,760K = 2(\overline{G}_T - L_TC_T)$, so the second integral reduces to

$$\frac{[\overline{G}_T - C_T(2L_T - 1)]^{a+1}}{a+1} + \frac{(C_T - \overline{G}_T)^{a+2} - [\overline{G}_T - C_T(2L_T - 1)]^{a+2}}{2(\overline{G}_T - L_TC_T)(a+1)(a+2)}.$$
$$\text{(V-17A)}$$

Adding the two integrals, we get for the numerator

$$\frac{8,760H_0}{C_0{}^a} \left[\frac{C_T{}^{a+1}}{a+1} - \frac{[\overline{G}_T - C_T(2L_T - 1)]^{a+2} - (C_T - \overline{G}_T)^{a+2}}{2(\overline{G}_T - L_TC_T)(a+1)(a+2)} \right].$$
$$\text{(V-18A)}$$

And dividing by $8,760 \, L_TC_T$, we get

$$\overline{H}_T = \frac{H_0}{(a+1) \, L_TC_TC_0{}^a} \left\{ C_T{}^{a+1} \right.$$

$$\left. - \frac{[\overline{G}_T - C_T(2L_T - 1)]^{a+2} - (C_T - \overline{G}_T)^{a+2}}{2(\overline{G}_T - L_TC_T)(a+2)} \right\}. \qquad \text{(V-19A)}$$

Note that

$$\left(\frac{C_T}{C_0} \right)^a = \left(e^{dT} \right)^{b/d} = e^{bT}, \quad \text{and} \quad e^{bT} = H_T/H_0. \quad \text{(V-20A)}$$

If we multiply V-19A by C_T^{a+1} and simultaneously divide inside the brackets by the same amount, we get

$$\overline{H}_T = \frac{H_0}{(a+1)\,L_T} \left(\frac{C_T}{C_0} \right)^a$$

$$\left\{ 1 - \frac{C_T\left(\dfrac{\overline{G}_T}{C_T} - 2L_T + 1\right)^{a+2} - C_T\left(1 - \dfrac{\overline{G}_T}{C_T}\right)^{a+2}}{2(\overline{G}_T - L_T C_T)(a+2)} \right\} \quad \text{(V-21A)}$$

$$= \frac{H_T}{(a+1)\,L_T} \left[1 - \frac{\left(\dfrac{\overline{G}_T}{C_T} - 2L_T + 1\right)^{a+2} - \left(1 - \dfrac{\overline{G}_T}{C_T}\right)^{a+2}}{2\left(\dfrac{\overline{G}_T}{C_T} - L_T\right)(a+2)} \right].$$

$$\text{(V-22A)}$$

Setting $S_T = \dfrac{\overline{G}_T}{C_T}$, we have

$$\overline{H}_T = \frac{H_T}{(a+1)\,L_T} \left[1 - \frac{(S_T - 2L_T + 1)^{a+2} - (1 - S_T)^{a+2}}{2(S_T - L_T)(a+2)} \right].$$

$$\text{(V-23A)},$$
$$\text{or equation (V-1)}$$

The term in the brackets will always be less than one except when the system operates at peak load all year-round ($L_T = S_T$).

Chapter VI

The Economics of Water Use

In the preceding chapter we presented a technique for determining the expected average thermal efficiency of the U.S. steam electric generating system in the year 1980. The derivation of this figure yields a measure of the utility industry's prospective demand for cooling services in that year. Having already indicated that this demand for cooling can hypothetically be satisfied in a number of different ways, we now turn to the investigation of the implications of this cooling load for U.S. water resources over the next two decades.

To translate cooling demand into a comparable demand for water requires a study of the economic environment in which utility managements must make their decisions about water use. But economic science is only a tool for deciding among more and less efficient techniques for achieving certain stated goals, and since the available techniques are limited by the state of scientific knowledge, a few more thermodynamic principles should be examined in preparation for an adequate economic survey.

THE PRINCIPLES OF COOLING

Up to now we have stressed the absolute magnitude (in Btu's) of the heat that must be eliminated. Nevertheless, each of the cycles we represented in Chapter III implies not only an output of waste heat but also an exit temperature for that heat. Furthermore, in order to cool any physical substance, we must not only find another material to absorb the heat, but this coolant must be at a lower temperature than the material from which heat is to be removed. Since most U.S. steam power plants discharge their condensed steam at a design temperature in the range of 80° to 90°F., a suitable coolant must therefore be a material which is available in large quantities under natural conditions and without excessive seasonality at temperatures sufficiently lower than 80°F. to permit transfer of heat at reasonably rapid rate. This rather stringent requirement immediately rules out of consideration for use in steam plants certain forms of cooling which are feasible in several other industrial processes.[1]

The principle that the coolant must be at a lower temperature than the exhaust steam and final condensate has two important corollaries. While heat transfer takes place as long as the two substances are at different temperatures, the transfer process tends to raise the lower temperature and, unless steam and condensate are present together, to reduce the upper one. Therefore our first corollary is that the substance which will absorb the greater quantity of heat per pound while rising 1°F. (i.e., the substance with the greater specific heat) is the "better" coolant from the standpoint of quantity required. The second conclusion we can draw from this principle is that, other things being equal, the closer the approach of the exit coolant temperature to that of the steam and condensate, the greater is the amount of heat that will be transferred per unit quantity of coolant. However, the closer the approach of the temperatures of the two streams, the slower the rate of heat transfer between them and the more heat transfer surface required in the condenser for effecting a particular condensation rate.

The selection of a suitable coolant for a particular heat transfer requirement is dictated by numerous factors. In addition to temperature and specific heat, the properties of viscosity and thermal conductivity have major influence on the heat transfer rate capabilities of the equipment. Convenient availability, low cost, and compatibility with metals are essential characteristics of useful coolants. Except for some highly specialized cooling applications, the only materials possessing the above requirements in sufficient degree for practical use are water and air.

In addition to its nearly ideal combination of properties as a coolant, water would have a unique additional advantage in power plant cooling if it were feasible to mix it directly with the exhaust steam, thereby approaching minimum coolant use and eliminating the cost of large

---

[1] Of course, this design temperature is not an immutable constant. If necessary it could be raised, on technological grounds, to a level which would permit the use of air cooling. Nevertheless, the economics of power production are such that the effort to achieve low exit temperatures is almost always more valuable than the increases in water costs it would entail. Cf. Chapter IV.

metal condensing surfaces. But because the steam used in modern generating stations is necessarily of a high purity, it is uneconomic to pollute it even with the minor quantities of dissolved salts in most raw water.[2] Once direct mixing is ruled out, it becomes necessary to invest in heat exchange equipment to condense the steam without direct contact, and the amount and cost of this capital then becomes vital to questions of cooling economy.

It is evident that only air is a reasonable competitor of water as far as coolant costs alone are concerned. There are, however, a number of very important objections to air. First, its temperature, while not appreciably different from water on a year-round average, is much more seasonal than water. This means that either exhaust steam temperatures would be much more variable if air were used, or that the cooling equipment must be designed for closer approaches in the hot summer months—at substantially greater capital costs. Furthermore, the physical properties of air are such that rates of heat transfer to it are much lower than to water. Consequently, even if great volumes of air were supplied, larger heat exchange surfaces would have to be provided than when water cooling is used. In general, these considerations rule out the use of air cooling of steam electric generating plants in the United States in the foreseeable future. In a few exceptional situations, air cooling of direct or indirect type may be introduced, but the effect on water demand cannot be expected to become appreciable.[3]

[2] Steam in a modern plant exceeds 99.9998 per cent purity, far greater than raw water (reference 54). Dissolved minerals must be removed from boiler feed water to prevent serious accumulation of mineral deposits in boiler tubes and to reduce corrosion. In the most modern super-critical plants, impurities may also enter the steam, and even minor impurities would seriously erode and corrode the turbine blades.

[3] An indirect, air-cooled, 120,000-kw steam-electric power station has been in service at Rugeley, England, since 1961 ("Cooling for Power Stations," by F. F. Ross and G. N. Stone; paper presented 25 February 1965 at Symposium on Industrial Cooling, Society of Chemical Industry, Great Britain). Steam leaving the turbine is condensed by direct contact with a spray of pure cold water in a jet condenser. A small portion of the resulting warm water is returned to the boiler, and the balance is cooled by circulation through banks of tubes in a large, natural-draft tower, where air rises up and around the heat exchanger tubes. The recooled water is pumped back to the jet condenser. In this normally cool climate (design conditions were 0.64 psi turbine exhaust pressure at 87°F. saturation temperature, 71°F. recooled water, and 52°F. atmospheric temperature), over-all thermal efficiency is nearly the same as a plant with a conventional wet-cooling tower. Although there are some operating problems associated with this new system, its feasibility under suitable conditions has been proved. Future prospects depend largely on the costs of transporting fuel, cooling water, and power relative to an additional capital investment in the air cooling system of about $15.00 per kilowatt of capacity.

Since air cooling is not competitive, economics dictates that water, at any foreseeable cost, must be the coolant used in steam plants. Should water become scarce, there is no alternative material, other than air, that could replace it, for even if an intermediate heat exchange medium were used in the condensers, economic efficiency would require that it be reused through the ultimate discard of its heat to water or to the atmosphere. The prospective demands for cooling steam electric plants must therefore be satisfied within the framework of water economics.

There are two theoretical possibilities for economizing on water use. The first would increase the amount of heat transferred to each gallon of water. Since the specific heat of water remains essentially constant over the range of temperatures involved in the cooling process, this can be accomplished by narrowing the differences between the temperature of the exhaust steam and the temperature of the cooling water as it leaves the condenser. Since this closer "approach," as it is called, involves constructing a larger, more costly condenser, the economy of this technique rests on a comparison of the reduced water costs and the larger capital outlays.

The second economizing technique is to reuse the cooling water several times instead of returning it to the aquifer after it once passes through the condenser. Since the water leaving the condenser is "polluted" by substantial additions of heat, the heat must be removed before the water is suitable for such reuse. This is customarily accomplished by passing the warm condenser water through a cooling tower,[4] where it is cooled by evaporation. It will be recalled that water absorbs a large amount of heat (about 1,000 Btu per pound) in passing from the liquid to the gaseous stage. The cooling tower operates by blowing air up through a descending spray or "rain" of warm cooling water, evaporating a small portion of the water and lowering the temperature of the remainder. The lower limit of the temperature in this process is the wet-bulb temperature—the minimum attainable temperature of water evaporating naturally into air of specified humidity and temperature.[5] Since the wet-bulb temperature is always less than the air temperature (i.e., the dry-bulb temperature), the difference being greater the dryer

[4] Cooling can also be accomplished by storing the warm condenser water in a pond until it cools. While this technique is effective, it is very land-intensive, and at the usual price of land in industrial locations, it is not widely used.

[5] Specifically, the reading of a thermometer, the bulb of which is covered with a wetted wick and held in the moving air stream.

is the air, power plant condensers can be operated at lower temperature with water cooling and subsequent heat discard to air in cooling towers than with direct air cooling of condensers. [6]

The cost-saving advantages of the cooling tower must, however, be purchased at the expense of other capital and operating outlays. The cooling water must be pumped to the top of a tower so that it trickles down through a current of fan-blown air. The pumps and fans require power, the structure requires maintenance, and the costs of construction must be reimbursed with a return on the capital. Furthermore, not all the water is saved; part of it is evaporated in the cooling process and must be replaced. And finally, since evaporation concentrates the impurities that might foul and corrode the system, some treatment facilities are usually necessary. The decision to recycle requires a careful balancing of costs and gains. The nature of these benefits and returns is indicated in the section below, headed "The Economics of Cooling Towers."

Before spelling out the incentives and disincentives of recycling, one important point needs to be made. We have been and will be discussing recycling from the viewpoint of the company making the decision. It should be clear, however, that no amount of recycling reduces the total amount of heat to be disposed of; more intensive use may not reduce, and may even increase, the burden of costs on other water users. A closer approach in the condenser will probably result in somewhat higher evaporation losses in the used cooling water due to higher discharge temperature, although normal atmospheric cooling will usually prevent an increase in the total heat load or the temperature in the stream some distance from the plant. [7] The use of a cooling tower reduces the heat load in the stream near the plant, but it also reduces the water flow because of tower evaporation. If the reduction in heat load is important to nearby downstream users, this will yield important external economies, but if the gross flow is most vital, the social effect will be harmful. We will touch on this matter later in the chapter (pages 75–76)

    [6] This is because although warm cooling water can be mixed with the air, generally it is not economical to mix cooling water with exhaust steam because of the cost of purifying the complete boiler feed-water replacement.

    [7] This natural cooling process is essentially the same as that taking place in the cooling tower, involving water evaporation and (if the dry-bulb temperature of the air is lower than the water temperature) heat conduction into the air. Hence, total water loss from the stream, after equilibrium is reached, is nearly the same whether once-through cooling or recycle cooling is used. Only a small loss of water as spray or mist from the cooling tower and a small evaporation difference due to limited radiation heat transfer from the stream, not occurring in the cooling tower, can alter the equality of far down-stream flows.

and again in Chapter VII, where it will be explained how the methodology developed in this study can be used in addressing problems of this nature.

VARYING CONDENSER SIZE

Given the amount of waste heat to be rejected in the condenser, the amount of water needed for the heat disposal can be reduced if each unit of water is made to carry a heavier load. Since the specific heat of water remains essentially constant over the relevant range of temperatures, each gallon of incoming water must be raised to a temperature closer to that of the steam that is to be condensed; hence, the size of the condenser must be increased. The economics of this method of conservation depends upon a comparison of the larger capital costs of the condenser with the savings from the cost of water.

Unfortunately, the economics of improving water efficiency in this way are not promising, as can be seen by the following example, illustrated in Figures 14 and 15. Assume that the exhaust steam condenses at 80°F. and the cooling water enters at 60°F. and leaves at 72°F. Each pound of cooling water carries away 12 Btu. Each additional degree by which the temperature of the outlet cooling water is raised means an additional Btu per pound of water. But each additional Btu results in a smaller *percentage* improvement in water use efficiency, so that the gains of successive increases in temperature are less than proportional. Since higher outlet temperatures also result in lower *absolute* levels of water use, this means that gains from increasing outlet temperatures (for given steam and inlet water temperatures) decline very rapidly.

The costs of these improvements, on the other hand, tend to increase sharply. The amount of heat that can be transferred across any given area of condenser surface depends upon the temperature differential between the coolant and the steam. As the exit temperature of the water is raised, the average differential narrows sharply.[8] As a result, the lower temperature differential requires larger condenser surfaces per kilowatt and even larger proportional increases in the *marginal* costs of increasing outlet temperatures (Figure 15). Due to scale economies in the construction of condensers, condenser costs do not rise quite as sharply as size. Nevertheless, substantial increases in water costs would be required to encourage further water conservation by this technique.

---

[8] The average that is relevant for these calculations is the log mean difference rather than the simple average (reference 36).

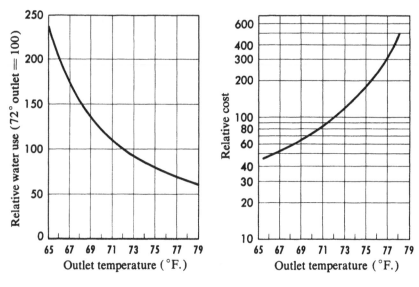

Figure 14. Relative water use for various outlet temperatures. (Inlet temperature = 60°F.; steam condensing temperature = 80°F.)

Figure 15. Relative costs of reducing water use by raising condenser outlet temperatures. (Inlet temperature = 60°F.; steam condensing temperature = 80°F.; cost of 72° outlet temperature = 100.)

If water should now cost enough to balance the costs and gains at an outlet temperature of 74°F. currently, a doubling of water costs would suffice to reduce water intake from 36½ to 31 gallons per kilowatt or by about 15 per cent, an arc elasticity of 0.15. Looked at from the other side, a 33 per cent reduction in condenser costs would permit only 9 per cent reduction in water use with economy.

While the responses to cost change are small, of themselves they are not negligible. We can, however, disregard these changes in making our projections, because any changes in the price of water large enough to affect condenser economics importantly are also large enough to encourage more radical methods of water conservation—particularly the use of cooling towers.

THE ECONOMICS OF COOLING TOWERS

The basic principles of cooling tower operation are simple, but the precise engineering properties are still a matter of some dispute. From an economic point of view, analysis is complicated by the large number of variables necessary for a complete solution of the optimizing problem. To make the problem manageable, it is necessary to simplify it by concentrating on those variables that are central to our purpose.

It will be recalled that cooling towers operate primarily by evaporating part of the circulating water. The advantage of cooling by evaporation derives from the latent heat of water. While only one Btu is required to raise the temperature of a pound of water 1°F., to evaporate that pound of water, about 1,000 Btu must be delivered to it from the rest of the water present (reference 37); thus, vaporization of one pound of water would cool 100 pounds by 10°F. The vaporizing and "losing" of one pound of water from 100 pounds causes enough cooling to permit recirculation of the remaining 99 pounds to the condenser for reuse. Thus, the withdrawal and fresh supply pumping requirement is only 1 per cent of the once-through demand. [9]

The evaporation technique by which the cooling tower operates has been described earlier in this chapter. In brief, air passed over water, vaporizes some of the water and cools the rest. [10] The air's ability to absorb water is a function of both the amount of vapor it already contains and the maximum it can hold, i.e., its relative humidity.

The wet-bulb temperature (defined on page 5) places an absolute lower limit on temperature reduction, but economic considerations necessarily raise this limit. As with the condenser, the smaller the temperature difference between the coolant and the cooled substance (or the "approach"), the more capital must be invested to achieve the cooling. Similarly, the greater the amount of water cooling required (i.e., the greater the temperature difference between the entering warm water and the cooled water produced, or the "range"), the larger must be the cool-

---

[9] For technical reasons, the actual withdrawal rate would be somewhat more than this rule of thumb indicates; other water losses account for the difference.

[10] A minor part of the heat is also dissipated from the water by non-evaporative cooling, i.e., if the water is warmer than the dry-bulb temperature of the air, as is usually the case, some of the heat is removed by simple heat transfer. The fraction of the heat lost in this way depends largely on the humidity and temperature of the air, but is generally small unless the air is cold or its relative humidity is high. Typically, less than 10 per cent of the heat is dissipated by sensible (non-evaporative) cooling. The effect of relative humidity on this proportion can be illustrated by comparing the evaporation occurring in a particular cooling tower when the wet-bulb temperature is 70°F. and the relative humidity is 50 per cent rather than 20 per cent. This increase in humidity results in an evaporation decrease of about 10 per cent, so if 1 pound were being evaporated for each 100 pounds cooled, the change in conditions would reduce evaporation to about 0.9 pound, the difference being accounted for by direct heat transfer to the air without evaporation (reference 33).

ing tower. Operating costs, largely the costs of pumping the air and the water through the tower, are similarly affected. These three variables, wet-bulb temperature, range, and approach, plus the costs of water, the energy for pumping water and air, and capital costs, are the central factors in the economics of cooling towers.[11]

Two of these three variables, however, are importantly affected by economic considerations outside of the cooling tower system. The range of cooling in the cooling tower must correspond to the temperature rise of the cooling water in the condenser in order to maintain a closed cycle. In addition, the temperature at which steam from the turbine is condensed is part of the basic steam cycle and is determined in part by the economics of fuel use. This condensing temperature of the exhaust steam plus the size of condenser determine the outlet temperature of the cooling water.[12] With the range and the cooling water outlet temperature taken as fixed, the approach used in the cooling tower is determined.[13] With capacity, approach, and range fixed, cooling tower capital and operating costs will vary only with the regional wet-bulb temperature.

To take the condenser range and approach and the exhaust steam temperature as parameters requires a little more justification. We have already suggested that for realistic water and fuel costs, the economics of the heat cycle are only negligibly affected by changes in water costs. To be sure, the *temperature* of the available cooling water is a vital factor in setting exhaust temperature, but water costs will not be important

[11] There are some economies of scale in cooling towers, especially in smaller stations, which make capacity a factor in these small units. However, for units of the size used in steam stations, we can ignore this added complication. One industry source shows quotations which indicate constant costs per gallon per minute of capacity for units larger than 15,000 to 20,000 gpm. This is roughly the size tower that would be used with a 40,000-kilowatt power station (reference 43).

[12] The optimum cooling water temperature also depends, to a minor extent, on cooling tower economics. There are trade-offs between condenser size, which is minimized by designing for a low exit cooling water temperature, and cooling tower size, which increases as this water temperature is reduced. The principal determinant of the ideal condenser outlet temperature is, however, the water flow rate and condenser size at which the total of capital costs and the present value of pumping costs is a minimum. In other words, total cooling costs comprising those of condenser and tower are considerably less sensitive to cooling tower design than to condenser design.

[13] Let T_i be the temperature of the inlet cooling water to the condenser, T_o be the temperature of the outlet water, T_s be the exhaust steam temperature and T_w be the wet-bulb temperature. Then $T_o - T_i$ must be the range of both the condenser and the tower. If T_s is determined by the economics of the steam cycle and $T_s - T_o$ by the economics of the condenser, they jointly determine T_o. If the range in the condenser is also fixed by condenser economics, then T_o and $T_o - T_i$ jointly determine T_i, and for any given wet-bulb temperature T_w, the cooling tower approach, $T_i - T_w$ is determined. T_w then becomes the central variable for the comparison of cooling tower costs with the cost of water.

in changing that temperature. If water withdrawal costs rise substantially, it is likely that by minimizing windage and blowdown losses wider and more effective use of cooling towers will yield water cost savings well over 90 per cent of once-through water cost[14] and may be a more economical way of saving water than changing condenser size, approach, or range (reference 43).

Cooling tower capital costs. A typical cooling tower consists of a tall rectangular wooden shell with internal wood grids or slats[15] over which the water trickles and falls in small streams or drops. Air is drawn into the tower through large openings near the base and up through the descending rain of drops, to exhaust fans at the top. Also common is a design employing forced draft fans near the tower base. The structure is basically simple and is not expected to undergo major design changes. Another type of tower, which employs natural draft rather than forced air supply, is not widely used in this country, mainly because the conditions of low design wet-bulb temperature and high relative humidity, at which these towers are best used, are not usually encountered here during the summer season when power station loads may be highest. Redwood which is the heart of the tower is costly and subject to slow deterioration, so there has been some substitution of plastic materials. Construction labor is an important element in costs. In all probability, technical improvements in tower construction will do no more than offset rising labor and material costs and no secular trend in tower costs is assumed.

While cooling towers of the size used by steam power plants do not display significant economies of scale, they do vary in cost depending on the difficulty of the cooling load placed on them (reference 43). The measure of this difficulty, commonly called the "relative rating factor" (or sometimes simply the "cooling factor"), is roughly proportional to total cost of the cooling tower at a given water flow rate and wet-bulb temperature. The larger the range (R), the higher is the relative rating factor (K) and the greater the tower cost. The larger the approach (A), on the other hand, the lower are the factor and the capital cost. For example, take as a base a tower designed to cool water from 90°F. to 80°F. when the wet-bulb temperature is 70°F.; that is, a tower with an R of 10° and an A of 10°. Arbitrarily, we assign a relative rating factor of 1.0 to this tower. Then if we wish to build a second tower with a range

[14] Because of the need for more effective water treatment when operating at lower blowdown rates, operating costs do not decline as much as gallonage.

[15] Because of tendency of the water to attack the wood, the lumber is carefully selected and expensive—mostly treated redwood heartwood of strict specifications.

of 10° but an approach of only 8°, we can find from design charts that the tower has a relative rating factor of 1.2 to 1.3, depending on its water pumping head, the higher factor applying to the shorter tower. Capital cost increases almost in proportion to the factor, so the second tower would cost roughly one-fourth more than the first. A good approximation for typical design towers is that with all other things constant, tower cost is inversely proportional to the approach.

Relative rating factors for a wide range of operating conditions are presented in Figures 16, 17, and 18.[16] In addition to their use in making preliminary designs and estimating costs, these graphs can assist in projecting cooling tower usage in the future. The first step requires a knowledge of the present average cooling water temperature, the usual cooling range, and the prevailing "design" wet-bulb temperature[17] in the region in question. These data may then be used with the three graphs to obtain the relative rating factor for the cooling tower.

In order to determine the capital cost of the tower we must then multiply this relative rating factor by the capital cost of the base unit. Estimates supplied by several industry sources indicate moderate variation in the cost of a large cooling tower operating at the base conditions (relative rating factor 1.0), and for the purposes of this analysis, a rough average of \$5.00 per gallon per minute (gpm) of capacity, exclusive of water basin, wiring, piping and circulating lines has been selected. The Marley Company, a leading manufacturer, has suggested that this additional equipment might raise the costs by about 60 per cent. Under that assumption, total capital cost for a tower will be about \$8.00 per gpm times the relative rating factor from Figures 16, 17, and 18. Thus, the complete system for cooling 50,000 gpm from 90° to 75°, at a wet-bulb temperature of 65°, has a relative rating factor (Figure 17) of 1.6 and an approximate cost of 8 × 1.6 × 50,000 = \$640,000.

Since our formula for the demand for efficiency involves water costs rather than tower costs, these capital costs must be apportioned over

[16] It is important to stress that these factors are per gpm rather than per kilowatt. While increasing the range (for any given approach) increases the capital cost per gpm, a smaller water flow rate is required to handle any given heat load, so that costs per kilowatt remain roughly constant. It is, however, simpler to evaluate the towers in units of water capacity.

Data from which Figures 16, 17, and 18 were constructed come from "Water Use and Related Costs with Cooling Towers" (reference 9), based on relations developed by The Marley Company.

[17] The appropriate ("design") wet-bulb temperature is customarily taken to be the temperature which is exceeded only 5 per cent of the time. Where the cooling tower is designed to compensate for low water flows in summer months only, somewhat more stringent temperature requirements may be necessary.

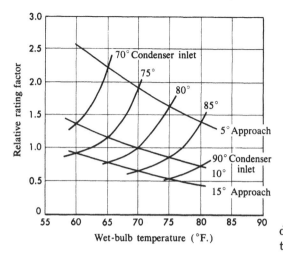

Figure 16. Cooling tower design for a water temperature range of 10°F.

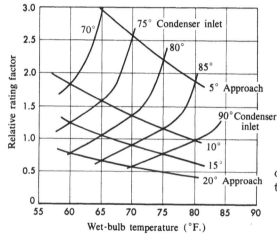

Figure 17. Cooling tower design for a water temperature range of 15°F.

the water supplied by the tower. The capital cost can be converted to the equivalent cost (in cents per 1,000 gallons) by multiplying it by

$$\frac{(r + 0.03 + t)}{5.256n} \quad {}^{18} \qquad (\text{VI-1})$$

---

[18] Cooling towers have long service life, estimated to range from twenty to over thirty years. In this analysis, large towers are estimated to last for thirty-three years, so the factor $(r + 0.03)$ reflects the costs of capital plus the cost of replacing capital. The denominator is equal to 525,600 minutes/year $\times$ n fraction of year in operation $\times$ 1/1,000 to convert into one-thousand gallon units and $\times$ 1/100 to convert dollars to cents.

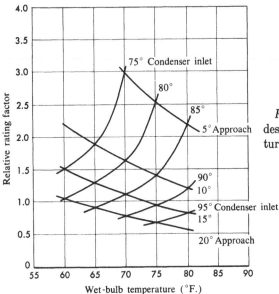

Figure 18. Cooling tower design for a water temperature range of 20°F.

where r = the annual cost of capital, t is property taxation rate, and n = load factor (all expressed as decimals).[19]

Cooling tower operating costs.[20] Operating costs for cooling towers fall into three major categories:

1) water to replace losses due to evaporation, windage, and discard of impurities (blowdown),[21]
2) chemicals to inhibit scale deposits and corrosion,
3) power costs for operating the fans and pumps in the tower.

The actual determination of these costs is a relatively complicated engineering problem. It is, however, possible to derive simplifying formulas which will provide reasonable estimates of recycling costs.

[19] Since cooling towers are constructed to serve plants rather than units, they are less likely to have secularly declining load factors and there is little reason to adjust the formula for that possibility.

[20] Much of the material in this section is drawn from Berg, Lane, and Larson, in the *Journal of the American Water Works Association*, March, 1964 (reference 9). The reader is referred to that journal for further reference, but is warned that much of the material applies to plants of smaller size than those used by steam power plants. The material from Berg, Lane, and Larson was supplemented by industry sources.

[21] Windage or drift loss of mist and spray is actually part of the total liquid water which must be discarded to avoid accumulation of impurities.

The first step in the determination of operating costs is the calculation of make-up water requirements, in physical terms. These costs depend upon the cooling range (R) and the number of times that the impurities in the water can be concentrated (C) while continuous evaporation occurs in the tower. This latter parameter depends in large part on the initial mineral content of the water. The ratio of make-up water to water flow through the towers measures the reduction in water withdrawals achieved by the use of cooling towers, although it should be remembered that in the case of recirculation, almost all the withdrawals are lost to the aquifer from which they are withdrawn, while in the case of once-through use, all withdrawals are returned.[22]

The formula relating these factors is

$$m = \frac{M}{L} = 0.001R\,\frac{C}{C-1} \qquad \text{(VI-2)}$$

where

m = the fraction of total water flow that must be made up by new water additions to compensate for evaporation and discard (blowdown)

M = make-up water requirements, gallons per minute

R = range (°F.)

C = number of times original impurities are concentrated

L = total water flow, gallons per minute.

The number of concentrations depends, as we have said, on original water purity and on the care with which operation is controlled. While it may be common practice to operate towers at three to four concentrations, power plant cooling should be more efficient, with values of C as high as eight to ten concentrations (reference 39). If we choose the midpoint of that range for gross estimates, our formula reduces to

$$m = 0.00113R. \qquad \text{(VI-3)}$$

This implies, for example, that for a cooling tower with a 20°F. range, $m = 0.0226$ and recirculation will reduce water withdrawals by 97.7 per cent of the once-through requirements.

The make-up water used in the tower must be treated in order to be

---

[22] The return of warmed water to a stream does not, however, greatly reduce the *over-all* evaporation resulting from power generation, as explained in the section of this chapter headed "Cooling Water Evaporative Losses and Minimum Requirements."

used most efficiently. It is usually treated with acid in order to prevent
scale formation, and with small amounts of other chemicals to inhibit
corrosion. The acid addition also extends the useful life of the wood in
the cooling tower. One common method of treatment entails costs that
can be estimated by using the following formula (in cents per 1,000
gallons of water treated):[23]

$$0.033Y + \frac{17}{C} \qquad\qquad \text{(VI-4)}$$

where Y is the alkalinity (as $CaCO_3$) in parts per million. For general
estimating purposes, one might take $Y = 150$ and $C = 9$ and derive
treatment costs for make-up water at 6.9 cents per 1,000 gallons
(reference 39).

   To find the total water costs for a cooling tower system, we must combine these results. For each 1,000 gallons of water circulated, $1,000m$ or
$1.13R$ gallons of water must be added as make-up. For each 1,000 gallons
of make-up we must pay out 6.9 cents for treatment, plus whatever costs
of water acquisition are involved. If water acquisition costs per 1,000
gallons are denoted by Wa cents, our estimating formula for water costs
of cooling tower operation (cents per 1,000 gallons of water circulated):

$$0.00113R\,(6.9 + Wa). \qquad\qquad \text{(VI-5)}$$

   The other major operating cost for cooling towers is the cost of power
for operating the pumps and fans associated with the tower. The power
required for fans varies with the relative rating factor and the rate of
water circulation. Using the same relative rating factor (K) we used to
analyze the capital costs of cooling towers, we find that fan horsepower
in large cooling towers may range from about $0.007KL$ to nearly double
that value. A typical tower is assumed here to require $0.01KL$ horsepower (reference 33). If we transform this into kilowatt-hours per 1,000
gallons circulated, by assuming a motor efficiency of 90 per cent, we
obtain

$$\frac{0.746\,(1000)\,(0.01)\,KL}{(0.90)\,60L} = 0.14\,K\,. \qquad\qquad \text{(VI-6)}$$

[23] The coefficients are based on a sulfuric acid price of 4 cents per pound. If large
quantities and low freight rates result in a 2-cent price, the formula would be
$0.017Y + (13.3/C)$; intermediate values would be proportional.

If power costs are p cents per kwh, then fan power costs are

$$0.14\,Kp\,.$$

Pumping costs are proportional to head (A) over which the water is pumped. A rule of thumb that can be used is that 0.5 kwh is consumed in lifting 1,000 gallons 100 feet, so that costs (in cents per 1,000 gallons pumped) will be $0.005\,Ap$. As a result, total power costs are (in cents per 1,000 gallons pumped)

$$(0.14K + 0.005A)p\,. \tag{VI-7}$$

By adding the costs of power and of make-up water and its treatment, we obtain the total operating cost of the cooling tower,

$$0.00113R(6.9 + Wa) + (0.14K + 0.005A)p\,, \tag{VI-8}$$

expressed as cents per 1,000 gallons circulated.

Steam cycle correction. Adding the capital and operating costs per 1,000 gallons (by use of formulas VI-1 and VI-8), we have the cost of supplying cooling water through recycling. Before this cost can be compared with the price of water in the region, however, we must adjust for the possible effect of recycling on the steam cycle itself.

Under conditions of negligible thermal pollution in streams, there is a rough correspondence between air and water temperatures, the latter averaging a few degrees below air temperature and a few degrees above the wet-bulb temperature. With wet-bulb temperatures normally lower than ambient air temperatures, it can usually be assumed that, allowing for the approach in the cooling tower, there is no important difference in the temperature of stream water and cooling tower effluent water in a given location. If, however, thermal pollution of streams is substantial or, on the contrary, the region's water supply comes from a cold body of water supplied from outside the region (like the ocean or the Great Lakes), temperature differences may become important. If a stream is polluted by heat, a cooling tower may be able to permit lower design steam exhaust pressure, improving thermal efficiency and reducing fuel costs. If the water source is unusually cold and of adequate supply, a cooling tower would not be economical, both because of its own cost and because the higher water temperature would reduce thermal efficiency and increase fuel costs.

There is some reason to expect a gradual rise in the temperature of many stream sources of water supply as future use of water for cooling becomes more intensive. Increased industrial activity may also leave some imprint on air temperature levels, but the effect of this is unlikely to be great enough or clear enough to warrant taking it into account when projecting future water use on a national level. In many industrial areas, however, thermal pollution of streams is already a problem and it is desirable, therefore, to have some means of making allowance for it, especially where water pollution legislation specifying heat as a pollutant is in effect or is likely to be proposed. Conversely, the planner would not want to overestimate the likelihood of recycling in areas with access to ample supplies of exceptionally cold water.

We have already derived (equation IV-5) the pounds of water required in the condenser for each kilowatt-hour produced:

$$\frac{3,413(1 - \beta\eta)}{\beta\eta\Delta T} = \frac{3,413(\alpha - H)}{H\Delta T} . \qquad \text{(VI-9)}$$

Conversely,

$$\frac{H\Delta T}{3,413(\alpha - H)} \qquad \text{(VI-10)}$$

is the number of kilowatt-hours produced for each pound of water pumped through the condenser, and

$$\frac{8,330\ H\Delta T}{3,413(\alpha - H)} \qquad \text{(VI-11)}$$

is the number of kilowatt-hours per 1,000 gallons. The fuel cost of a kilowatt-hour is (equation IV-3)

$$\frac{0.003413f}{H} \qquad \text{(VI-12)}$$

so that the fuel cost (in cents) of the power generated per 1,000 gallons of water used is

$$\frac{8,330H\Delta T}{3,413(\alpha - H)} \times \frac{0.003413f}{H} = \frac{0.00833f\Delta T}{\alpha - H} . \qquad \text{(VI-13)}$$

What we now need is a measure of the increase (decrease) in generation made possible by lowering (increasing) inlet water temperature.

As is indicated in Appendix A, the efficiency of a Carnot cycle can be calculated by the formula

$$\frac{T_2 - T_1}{T_2} \qquad \text{(VI-14)}$$

where T_2 is the upper temperature limit of the cycle and T_1 is the lower limit, both temperatures being expressed on the absolute scale. For the ordinary steam cycle, this must be modified by substituting T_A, the average temperature of heat input, for T_2, so that

$$\eta = \frac{T_A - T_1}{T_A} . \qquad \text{(VI-15)}$$

Given this,

$$\frac{d\eta}{dT_1} = -\frac{1}{T_A} . \qquad \text{(VI-16)}$$

If we rearrange equation VI-15, we get

$$T_A = \frac{T_1}{1 - \eta}$$

$$\frac{d\eta}{dT_1} = -\frac{(1 - \eta)}{T_1} . \qquad \text{(VI-17)}$$

Now, equation VI-13 above gives the fuel cost of the kilowatt-hours produced when 1,000 gallons of cooling water are used. If water inlet temperature is reduced by dT_1, the increase in efficiency will mean that the same fuel costs will result in a fractional increase in kilowatt-hours produced of

$$\frac{d\eta}{\eta} = -\frac{(1 - \eta)}{\eta T_1} dT_1 \qquad \text{(VI-18)}$$

or conversely, producing the same number of kilowatt-hours will reduce fuel costs by the fraction $d\eta/\eta$. This saving is

$$\frac{0.00833f}{\alpha - H} \Delta T \times \frac{d\eta}{\eta} = -\frac{0.00833f(1 - \eta) \Delta T \, dT_1}{(\alpha - H) T_1 \eta} \qquad \text{(VI-19)}$$

cents per 1,000 gallons of water used. (If dT_1 is positive, there is a negative saving.)

In the short run, this saving in operating cost represents the full advantage of a decrease in cooling water temperature. That is, given the level of capacity, a reduction in the temperature of the inlet condenser water will imply a reduction in the costs of serving the given load.[24] In the longer run, however, when capacity is permitted to vary, the true effect is to reduce the amount of additional investment required to meet the expanding load. In other words, given the present level of operating costs, a decrease in water temperature is equivalent to a fractional increase $d\eta/\eta$ in capacity. In this case, the savings per kilowatt of present capacity can be computed to be

$$\frac{d\eta}{\eta} \times \text{cost of a new unit of capacity per kilowatt} \qquad \text{(VI-20)}$$

and the reduction in costs per 1,000 gallons of recirculated water can be computed by dividing the savings (VI-20) by the number of 1,000 gallons used per kilowatt,[25]

$$\frac{3,600 n_0 (1 - \beta\eta)(1 - e^{-\nu t})}{\beta\eta \Delta T y} . \qquad \text{(VI-21)}$$

Therefore, the water cost reduction (in *cents* per 1,000 gallons used) is equal to the cost (in dollars per kilowatt) of a unit of new capacity multiplied by

$$\frac{d\eta}{\eta} \frac{\beta\eta y \Delta T \times 100}{3,600 n_0 (1 - \beta\eta)(1 - e^{-\nu t})} = \frac{d\eta}{\eta} \frac{\beta\eta y \Delta T}{36 n_0 (1 - \beta\eta)(1 - e^{-\nu t})}$$

$$\text{(VI-22)}$$

[24] However, with an existing turbine designed for a certain exhaust pressure, a very large decrease in temperature of the cooling water supply to the condenser may reverse this efficiency increase by causing excessive steam velocities and wasted energy in the low-pressure stage of the turbine. If this should be a problem in a particular power plant, simple means for avoiding excessively low cooling water temperatures can readily be provided.

[25] Obtained from equation IV-8, as follows:

$$\text{Water costs} = \frac{36 \, wn \, (1 - \beta\eta)}{\beta\eta \Delta T} \text{ \$/kw-yr.}$$

Dividing by water cost, $W/100$ \$ per 1,000 gal,

$$\text{Water use} = \frac{3,600 \, n \, (1 - \beta\eta)}{\beta\eta \Delta T} \text{ 1,000 gal/kw-yr.}$$

If $n = n_0 e^{-\nu t}$, water use over life of the plant is

$$\int_0^\tau \frac{3,600 \, n_0 \, (1 - \beta\eta) e^{-\nu t}}{\beta\eta \Delta T} dt \text{ 1,000 gal/kw}$$

Integrating, $\dfrac{3,600 \, n_0 \, (1 - \beta\eta) \, (1 - e^{-\nu t})}{\beta\eta \Delta T y}$ 1,000 gal/kw.

and substituting from equation VI-18

$$\frac{-(1 - \eta) \beta y \, \Delta T \, dT_1}{36n_o(1 - \beta\eta)(1 - e^{-\nu t}) \, T_1} \tag{VI-23}$$

times cost (dollars) of 1 kw of new capacity = water savings in cents per 1,000 gallons.

If the cooling water temperature at condenser inlet differs (because of cooling tower use) from that of the natural supply in a once-through system, the costs of recirculated cooling water determined by use of capital and operating cost formulas may now be adjusted by applying formula VI-23. Substitution of this temperature difference (dT_1) and appropriate values for the other terms in formula VI-23 results in a figure which is subtracted from the total water costs obtained from formulas VI-1 and VI-8 (or added if the recirculated water is warmer than the natural supply).

COOLING WATER EVAPORATIVE LOSSES AND MINIMUM REQUIREMENTS

It will be recalled that, except for heat losses directly to the atmosphere in the form of hot flue gases from the boiler furnance, essentially all the energy in the fuel not converted to electricity is transferred to cooling water. It will also be recalled that the principal process by which this unusable energy ultimately enters the atmosphere is the evaporation of some of the cooling water, whether in a cooling tower or from the surface of a stream into which warm water is discharged. These facts permit an important conclusion as to net water use, or disappearance of water, in steam-electric power generation.

A quantitative relationship between water disappearance and total generation may be obtained by a method similar to that used for total once-through use, equation IV-5. Instead of ΔT, which represents heat removal per pound of circulated water, we substitute 1,050 Btu removed per pound of *evaporated* water, whence,

$$\frac{\text{pounds evaporated}}{\text{kwh}} = \frac{3,413(1 - \beta\eta)}{1,050\beta\eta} = \frac{3.25(1 - \beta\eta)}{\beta\eta}. \tag{VI-24}$$

In terms of over-all thermal efficiency, H and boiler furnace efficiency, α (usually about 0.9),

$$\frac{\text{pounds evaporated}}{\text{kwh}} = 3.25 \left(\frac{\alpha}{H} - 1 \right), \text{ and} \qquad \text{(VI-25)}$$

$$\text{gallons evaporated/kwh} = 0.39 \left(\frac{\alpha}{H} - 1 \right). \qquad \text{(VI-26)}$$

Regardless of power plant capacity and efficiency, type of fuel, gross water supply rate, and the use or extent of recycling, approximately one pound of water evaporates into the atmosphere for each thousand Btu entering the cooling water. Allowing 10 per cent fuel energy loss in flue gases ($\alpha = 0.9$), this means that over the very broad range of power plant efficiencies from 25 per cent to 40 per cent, corresponding to heat rates of 13,600 Btu to 8,500 Btu per kilowatt-hour generated, the heat discarded to cooling water results in evaporation of about 8.8 pounds down to about 4.2 pounds of water per kilowatt-hour generated. The higher figure, slightly over one gallon per kilowatt-hour, represents a current practical maximum water consumption in old or small power plants; the lower figure, about one-half gallon per kilowatt-hour, is realized in the newest and largest plants, and represents very nearly a long-range minimum water requirement in steam power generation.

Although the computation of the above one-gallon and half-gallon figures does not include such minor factors as slight additional water losses to the atmosphere caused by spray and mist from cooling towers nor slight reduction in evaporation from streams because of heat radiation to the atmosphere, these averages are nevertheless useful in estimating minimum water supply requirements and net water consumption for electric power generation. The figures may be applied to individual plants, to regions, or to the country as a whole, simply by applying the pertinent plant efficiencies, or heat rates, to the total electrical output, in kilowatt-hours.

Thus, at a current average efficiency in U.S. central station power plants of about 33 per cent (reference 18), corresponding to a heat rate of 10,300 Btu per kilowatt-hour, approximately 0.7 gallon of water evaporates into the atmosphere per kilowatt-hour generated. A total generation of approximately 600 billion kilowatt-hours per year (reference 18) thus results in the evaporation of about 400 billion gallons of water. As average power plant efficiencies increase toward 40 per cent during the next two decades, minimum consumptive water requirements will decrease from the present 0.7 toward 0.5 gallon per kilowatt-hour, and practical estimates of net water use, or disappearance, can be based on these figures.

It is seen that water disappearance is a very small fraction of total water usage in power generation—only about 1 per cent of the national withdrawal average of about 50 gallons per kilowatt-hour (reference 24). It is also seen that the relationship between disappearance and total generation is very simple, about one-half to three-fourths gallon per kilowatt-hour generated. Projections of net water consumption by steam electric plants are therefore easily determined from electricity output estimates.

ALLOWANCE FOR POLLUTION

The optimum level of water quality in any water course depends upon both the costs imposed on successive users by waste discharge and the costs of curbing the waste discharge or otherwise affecting quality in the water course.[26] In the usual instance, the quality of effluent can be improved in a more or less continuous fashion with correspondingly rising costs. In the case of power plants, however, the economies of recirculation are such that if it is instituted at all, it will be done essentially completely and, accordingly, little or no heat will be discharged. Thus, if offsite costs (and other economies resulting from recirculation) are high enough to justify recirculation, the temperature of the water source will be unaffected by the use of the thermal electric plant and downstream users will suffer no heat damage. There will be, however, a small increase in the dissolved solids content of the stream due to the chemical treatment of the recirculated water and its ultimate discharge.

On the other hand, it might be much less costly, given the relative magnitudes of the costs of quality improvement and the potential damage to downstream users, to leave the water in its warmer state. One can even visualize an extreme case in which a downstream navigational user might be benefitted by heat pollution that reduced winter ice formation; such a user might not only prefer pollution to continue, but would be willing to pay for doing nothing to the water.

From an economic point of view, welfare might be increased most by permitting arrangements to be made in as inexpensive a manner as possible and considering all users as a whole.

This is not the place to deal extensively with the broader problems of water quality which are complicated enough to deserve separate study.

---

[26] By the augmentation of low flows through releases from reservoir storage, for example.

It is only necessary here to point out that, in essence, any attempted solution to the problem will depend upon its two aspects: the cost of quality improvement at points of effluent discharge, and the assessment of damages to successive users. This study contributes two pieces in the skein of information required.

Pollution of intake water affects an electric utility in three ways. First, the steam used in the thermal cycle must be very pure. Thus, if the water for steam has any non-thermal pollutants, it must be purified. These costs are significant, but this study does not deal with them. Second, if the condenser water carries non-thermal pollutants, the huge volumes of water required, at least in once-through procedures, may foul the condenser and cause more frequent interruptions for cleaning. These costs are not within the scope of the present work. Third, thermal pollution of the condenser water will reduce the fuel efficiency of the plant, raise costs, and lower capacity.

In this connection, the steam cycle correction equation (VI-23) offers a method for evaluating the loss to the utility. Since the formula measures the *marginal* approximate cost of a change in inlet water temperature, it also measures the cost of an increase in water temperature due to upstream heat input. It should be noted, however, that the cost to the utility depends upon the temperature change of its *inlet* water and not the change in the upstream waste discharger's outlet water. Dilution by the full stream flow and subsequent natural atmospheric cooling will usually cause these two temperatures to differ.

Within the limited scope of this study it is not possible for us to calculate the effect on downstream users of a power company's heat disposal. We can, however, estimate the cost to a utility of maintaining the outlet water at the same temperature as its inlet water. This, in short, is the cost to the utility of recycling its water and is equal to the amounts indicated in this chapter to be the capital and operating costs of cooling towers under the particular conditions facing the utility. Thus, the data developed in this chapter for the purpose of determining the volume of electric utility use also serve to answer some of the questions vital to an analysis of the economics of water pollution.

However, it must be appreciated that the costs of recirculation may not be *net* costs of abating heat discharge, for certain internal economies accrue to the plant as a result of recirculation and frequently these economies are great enough in themselves to justify the practice. In any specific instance, net costs must be determined by subtracting the internal economies.

COMPARISON OF COOLING SYSTEMS

The sum of cooling tower capital costs, operating costs, and the heat rate adjustment may be compared with the delivered cost of water from the local supply. If the sum of these costs is greater than that of local water at the power plant, operations should be on a once-through basis unless, of course, offsite costs due to heat discharge are high enough to justify recirculation when considered in conjunction with the possible reduction in intake costs. On the other hand, any region or part of a region where water is more expensive than the costs of recycling will find it advantageous to recycle. In the following chapter, we will indicate how these adjustments should be made.

It must be recognized that the above cost comparison is based on total costs of the two alternatives, for a specific plant only, and that the most economical alternative for one plant may represent a greater water expenditure for the region as a whole. For example, a low-cost cold water supply may be the cheapest source for one plant, so once-through use would be employed. The warm effluent, if immediately reused in power plant condensers a short distance downstream, could then result either in higher fuel costs in that plant or in added expense of cooling tower operation for water temperature reduction. If downstream plants are at adequate distance from upstream users, however, or if streamflows are many times the individual withdrawal rates, temperature effects on downstream users may be negligible. To minimize total cooling costs in a region—as on a particular river system—comparisons would have to be made for each plant and the effects of one plant on another would have to be considered. The combination of cooling water systems representing minimum total cost would then be the regional optimum if there are no other damaging effects associated with the discharge of the hot effluent; if there are, these costs would also have to be weighed in the balance.

Chapter VII

A Regional Model for Projections

In the previous six chapters, we have indicated the main factors that regulate the demand for water by steam electric power plants, and we have discussed in quantitative fashion the relationships between these factors and the use of water. The procedure may seem unduly abstract, but the level of abstraction is the price paid for generality. The formulas are in terms as general as is practical, primarily so that they can be used by water planners possessing little engineering or economic training.

This does not mean that the need for judgment or intuition has been eliminated. We think we have reduced the necessity for subjective judgment but, more important, we believe we have stated our procedures in such a way that the planner cannot only make an estimate, but can also calculate the sensitivity of this estimate to a change of assumption. If he assumes, subject to uncertainty, that the 1980 fuel price in his region will be 28 cents per million Btu, he can also estimate how far off the mark he will be if the rate turns out to be 25 or 35 cents.

To illustrate how the proposed method may be used, we now proceed to calculate expected water use in 1980 for a specific region. Since the purpose of this projection is to illustrate the model, it was desirable to find a region with diverse hydrologic characteristics. For this illustration, however, it was considered unnecessary to ascertain the exact characteristics of the region chosen; most of the data are, therefore, from the most readily available sources, and some of them, particularly the water cost information, are merely suggestive.

The region selected is the West South Central, comprising the four states of Arkansas, Louisiana, Oklahoma, and Texas. Hydrologically, the main river basin is the Arkansas-White-Red, with the Rio Grande watering the southern parts of the region. In the eastern portion, there is a good water supply, with the dry area generally west of a line drawn through Waco and Austin, Texas. It is a fast-growing area electrically, increasing its capacity 175 per cent from 1950 to 1958 (reference 18). This rate of growth is 35 per cent larger than in the country as a whole, due partly to rapid industrial development and partly to a high level of consumer use spurred by air conditioning. Almost all (90 per cent) of the region's capacity is thermal.

In making projections, the procedure is somewhat different from that developed in previous chapters. We need to know the cost of water to the utility before we can determine its demand for thermal efficiency, but we must first examine the utility's recycling opportunities before we can determine its water costs. The procedure can be outlined as follows:

1) Determine projected recycled and unrecycled water costs and set the utility's costs equal to the lower of these two,
2) Determine the projected cost of fuel,
3) Calculate the desired level of best thermal efficiency,
4) Obtain estimates of future electrical capacity and demand,
5) Estimate the average thermal efficiency in the year to be forecast,
6) Calculate forecast water use.

First, we need some facts about the region we have chosen. In 1954, the census of water use by steam electric generating plants, conducted by the Federal Power Commission, listed 79 gallons per kilowatt-hour as total use, including recirculated water, in the West South Central region (reference 24). Since the area heat rate in that year was 12,500 Btu per kilowatt-hour (reference 18), this implies that condenser practice in the region was to raise cooling water temperature an average of 12°F.[1] As has been explained in Chapter VI, this datum is unlikely to change in the next two decades, primarily because any increases in water costs would simply be met by greater use of recirculation rather than by reduction of flow through the condensers.

The costs of recirculation in the region can be estimated by means of the graphs and formulas presented in Chapter VI in conjunction with generally available meteorological data. Fortunately for the region, the wet-bulb temperature, which is vital to these calculations, is lowest in those areas in which water is most scarce, a factor which has undoubtedly encouraged growth of recycling in the region. In 1954, withdrawals were only 45 per cent of use, a rate which is the lowest in the nation outside

[1] This average includes both recirculation systems and once-through systems. A somewhat more realistic appraisal could be made by separating the water usage into the two types, and computing the temperature rise for each.

of the small steam generation in the Mountain States. Furthermore, almost none of this recycling was undertaken in Louisiana, which has ample water and a high summer wet-bulb temperature: 80°F. is exceeded 5 per cent of the time, as opposed to 70°F. in the western part of Texas and Oklahoma. (References 24; 42, p. 20.)

If we use these summer wet-bulbs in determining the design approach temperatures and 12°F. as the range, we can obtain the relative rating factor for cooling towers by interpolation from Figures 16 and 17. Costs can then be estimated.

If condenser inlet water temperatures are 85°F., the relative rating factor for a cooling tower in the arid western part of the region (70°F. design wet-bulb and R = 12°), will be about 0.7, while in the east (80°F. design wet-bulb), it will be about 1.65. At "standard" total costs of $8.00 per gpm, capital costs will be $5.60 and $13.20 per gpm, respectively. Use of formula VI-1, in conjunction with cost factors shown in Table 3, then yields annual capital costs of 2.4 mills per 1,000 gallons in the western section and 5.6 mills in the east. Operating costs per 1,000

TABLE 3. BASIC DATA FOR CALCULATING THE DEMAND CURVE FOR THERMAL EFFICIENCY IN THE WEST SOUTH CENTRAL REGION

	1962		1980	
Item	West South Central	National	West South Central	National
Thermal capacity (MW)	17,646	150,750	51,200	
Total capability (MW)		198,050		
Noncoincident peak (MW)		151,250	44,500	
Generation—total (million kwh)	76,876	943,053		
—steam (million kwh)	73,278	678,978	230,912	
Heat rate (Btu/kwh)	10,887	10,493		
Thermal efficiency (H) (%)	31.5	32.5		
Fuel cost (f) (¢/million Btu)	19.3	26.4		
Cost of capital (r) (as %)	6.0	6.0	6.0	6.0
Expected fuel price rise (x) (as %)	2.0		1.0	1.0
Expected load decay (y) (as %)	4.0		3.5	
Expected change in water costs (z)	0		0	
Load factor (n) (as %)	47.4		51.5	
Property taxation rate (t) (as %)	1.5			
Raw water cost (w) (¢/M gal.)	1.0			
Power cost for cooling tower fans and pumps (p) (¢/kwh)	1.0			

Source: Data from Edison Electric Institute, *Statistical Yearbook*, 1951-64; or estimated by authors.

gallons in the two regions (formula VI-8) will differ little—3.8 mills in the west and 5.1 in the east, assuming 1 cent power, 1 cent/1000 gallons raw water, 35-foot pumping head. Total water costs under these conditions then range from 0.62 cent per 1,000 gallons in the west to 1.07 cent per 1,000 gallons in Louisiana.

All of the water cost figures noted above are predicated on achieving, by recirculation, the water temperature which existed prior to recycling. On the other hand, utilities in a region with high water temperatures and low wet-bulbs can take advantage of recirculation to lower cooling water inlet temperatures and improve the heat cycle, and, in fact, they do.[2]

If we assume, as an illustration of this point, that cooling towers in the western portion of the West South Central region can provide water to power plant condensers at a temperature 5° lower than the natural supply, the adjustment formula VI-23[3] suggests that water costs should be credited by about 0.4 mill per 1,000 gallons, bringing total costs in the west down to 0.58 cent per 1,000 gallons. If recycling were used to lower water temperatures by 10°F., the heat cycle credit would be 0.8 mill per 1,000 gallons, but almost all of this credit would be counterbalanced by the increased capital costs of achieving the lower temperatures.

With all adjustments made, then, regional recycling costs will vary from 0.58 cent per 1,000 gallons in the western part of the region to 1.07 cent in Louisiana. For illustrative purposes, however, we assume that by 1980 the costs of unrecycled water will vary from over 2 cents per 1,000 gallons in the west to less than 1 cent in the east. We further assume that, as a result, all of the region west of the Waco-Austin line

[2] Ordinarily, natural surface water temperatures and wet-bulb temperatures parallel each other, the latter being a few degrees lower. However, high solar radiation, low wind velocity, and thermal pollution of streams can cause water temperatures to rise perhaps 20 degrees above wet-bulb temperature.

[3] Factors for substitution in formula VI-23:

$\beta\eta\alpha$ = 0.315 (over-all efficiency from Table 3)

α = 0.90 (assumed boiler efficiency)

β = 0.90 (assumed turbine-generator efficiency)

η = 0.389 (by calculation from preceeding factors)

y = 0.04 (load factor decay rate from Table 3)

ΔT = 12° (cooling tower typical range, note 1 of this chapter)

dT_1 = 5° (assumed reduction in cooling water temperature below natural supply)

M_o = 0.474 (load factor, Table 3)

t = 18 (number of years 1962 to 1980)

T_1 = 545 (assumed absolute temperature of cooling water to condensers)

$100/kw (assumed approximate capital cost of power plant)

will use recycling by 1980 and that the area to the east will do no re-circulation. Obviously, the regional demarcation line will be less clear-cut than this. It is quite possible, in fact, that different plants located relatively close together will vary with respect to recirculation because of local differences in water costs. For this illustration, however, it is practical to calculate on the basis of a sharp regional break.

An additional important point should be noted here. In our illustration, if the offsite (heat pollution) costs had risen to 0.24 cent or more in the east, combined water cost and offsite pollution cost would have justified recirculation. Again, the offsite costs are likely to differ appreciably from plant to plant—a circumstance which the water planner must take into account. We hope the way in which our methodology can be extended to these more detailed situations will be clear.

On our assumptions, then, average use (withdrawal) west of the line will be about 1.4 per cent of once-through use.[4] Use in the east will de-pend upon the average thermal efficiency of installed plants, which we must now calculate. For this purpose, we must first determine the ex-pected efficiency of "best" plants. In the model, this expected "best" efficiency is determined by our equation for the demand curve for thermal efficiency (equation IV-14) and the supply curve for thermal efficiency (Figure 6 with its notes). We proceed first to a calculation of the demand curve.

THE DEMAND CURVE FOR THERMAL EFFICIENCY IN THE WEST SOUTH CENTRAL REGION

The basic data for the calculation have been listed in Table 3. Here we present the underlying assumptions with, however, the reiterated warning that the projection is intended to be merely illustrative; thus, while realistic assumptions have been used, based on the work of others, it is in no sense our forecast of the future. Anyone undertaking such a forecast would want to determine, more carefully than our limited purpose required, these fundamental data.

Because of its location in the center of the United States petroleum and natural gas producing region, the West South Central has the lowest fuel cost of the country's ten regions.[5] In 1962, this cost was 19.3 cents

---

[4] Calculated from formula VI-3 for make-up requirements in Chapter VI. Thus, $m = 0.00113\ R$ and $R = 12°$, whence $m = 0.014$ or 1.4 per cent.

[5] These are the regions used by the Edison Electric Institute; they conform generally to the Census regions.

per million Btu as compared with a national average of 26.4 cents (reference 18). As a result, new plants in the region have generally ranked substantially higher in heat rate than plants of similar size built in other regions of the country. In terms of installed capacity this region has been one of the most rapidly growing sections of the country. Thermal capacity there doubled from 1957 to 1962, compared with the national increase of 51 per cent (reference 23). Because of this rapid growth, average efficiency in the region is close to "best" levels, and hence is greater than in some other regions with higher fuel costs but slower growth. Thus, while best efficiency is lower in the West South Central region, average efficiency is higher than in New England, which grew only one-third as fast and the Mountain States, which grew only half as fast (reference 18.)

Another important factor in making efficiency decisions for long-lived plants in the region is that while fuel costs have been lower than elsewhere in the United States, they have also been rising faster then elsewhere. Between 1957 and 1962, fuel costs in West South Central rose 42 per cent, while the national average fuel cost was falling 2.5 per cent. The rapid rise is due to the increase in demand for cheap fuel as plants moved into the region and as pipeline networks carried the fuel elsewhere in the nation.

By 1980, most of this adjustment will have been completed. A National Planning Association study, by Perry D. Teitelbaum (reference 61), estimates that while 1954 fuel prices in the region were only 42 per cent of the U.S. average, the 1980 figure will be 80 per cent of the predicted national figure, or 28 cents per million Btu in 1957 prices. [6]

The remaining discrepancy between regional and national price levels results largely from transport costs which are unlikely to be obliterated. We assume, therefore, that in 1980, the expected rate of increase in regional fuel prices over the ensuing years will be the same as that of the nation generally. This implies a rate of increase in fuel prices of about 2 per cent per year to 1980, and an increase of 1 per cent thereafter.

The West South Central region operates at a somewhat lower rate relative to generating capacity than the country as a whole. In 1962, its rate was only 47.4 per cent as compared with 51.4 per cent for the entire United States excluding Hawaii and Alaska (reference 18). Most

[6] The Teitelbaum data are for the Federal Power Commission region of which West South Central is a part. These have been adjusted to apply to the West South Central area.

of this difference may be accounted for by the very high summer air-conditioning peak, which is not balanced by a winter heating load as in the northern part of the nation. While the differential between the region and the nation should continue, it is likely that both figures should rise as the ratio of capability to peak load drops from its present 1.25 level to the ratio of 1.15, which seems to represent the asserted goal of utility executives. We have forecast a 1980 utilization rate for steam capacity in West South Central of about 50 per cent. This corresponds to a decay rate for the load factor of new plants (initially 100 per cent load factor) of about 4 per cent per year. At exactly 4 per cent decay, the load factor in 1980 would be 51.5 per cent.

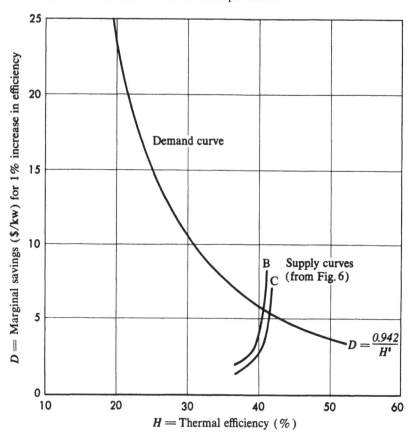

Figure 19. Determination of best power plant efficiency in 1980, Southwestern United States, as an illustrative problem.

Using these data, we can derive the curve of marginal fuel savings. Substituting these values in equation IV-14, we find that for a 1 per cent change in H the savings are:

$$D = \frac{1}{H^2}\left[\frac{0.003\,(19.3)[1 - e^{-35(0.080)}]}{0.080} + \frac{0.36\,(0.9)\,(1)[1 - e^{-35(0.10)}]}{0.10\,(12)}\right].$$

Comparing this equation with the supply curves for 1980—this can best be done by superimposing a graph of the above equation on the supply curve—we find that the "best" efficiency plant in 1980 will be about 41 per cent, equivalent to a net station heat rate of about 8,320 Btu per kilowatt-hour. (See Figure 19).[7]

To move from our projection of best efficiency to one for average efficiency, we now turn to equation V-1. We already have $H_T = 0.41$. Utility executives frequently report a "desired" ratio of reserve margin to peak load of 15 per cent; this implies a ratio of peak load to total capacity of 0.87.[8] Furthermore, since in 1962 best efficiency plants in the area were 34 per cent efficient, the average annual rate of improvement in efficiency works out to be 1.0 per cent. Capacity in 1962 in the region was 17,646,000 kilowatts. By 1980, it is projected at 51,200,000 kilowatts; a gain of 6.0 per cent per year. From these two growth rates, $a = b/d = 1.0/6.0 = 0.167$. Substituting these values into equation V-1, we get:

$$\overline{H}_T = \frac{0.795}{(a + 1)}\left[1 - \frac{(0.84)^{(a+2)} - (0.13)^{(a+2)}}{0.71\,(a + 2)}\right]$$

$$\overline{H}_T = \frac{0.795}{1.167}\left[1 - \frac{0.84^{2.167} - 0.13^{2.167}}{0.71\,(2.167)}\right]$$

yielding an average efficiency of 38.4 per cent.

---

[7] Note that what we have done is to draw the curve $D = 0.942/H^2$ on a graph (Figure 19) on which two supply curves for 1980, transposed from Figure 6, have also been plotted. The thermal efficiency at the intersection point is about 41 or 42 per cent, depending on which of the 1980 supply projections is used.

[8] Reserve margin is defined as the difference between capability and peak load. In our symbolism, this is $C_T - \overline{G}_T$. Therefore, a reserve ratio of $(C_T - \overline{G}_T)/\overline{G}_T = 0.15$ implies a ratio of $\overline{G}_T/C_T$ equal to $1/1.15$ or 0.87.

To find unrecycled water requirements, we turn back to equation IV-5. Assuming a boiler efficiency of 0.90, $\beta\eta = \bar{H}/\alpha = 42.7$ per cent. Then substituting this value of $\beta\eta$ and $\Delta T = 12$, we get:

$$\frac{3,413(1 - \beta\eta)}{8.33\beta\eta\Delta T} = 46 \text{ gallons/kwh.}$$

This is the rate of use we would expect in the eastern part of the region. The withdrawal rate in the western part would be about 1.4 per cent of that, or 0.8 gallon per kilowatt-hour.

Appendix A

The Physics of Thermal Efficiency

In the present state of power technology, the most efficient machine on the drawing boards for converting fuel energy into electric energy, using steam as an intermediary has an efficiency of only 42 per cent. However, while there will undoubtedly be further improvement, the use of percentages to indicate thermal efficiencies can be seriously misleading if one assumes that the large difference between this figure and 100 per cent represents great potential fuel savings in generating electricity. All machines operate, of course, at less than 100 per cent efficiency. Under laboratory conditions, however, when economic limitations can often be avoided, careful attention to precise construction can produce mechanical and electrical machinery capable of approaching the theoretical 100 per cent limit, and with the improvement of techniques for building close tolerances into machines, constant improvement has been achieved even in the commercial field. In this manner, for example, the transformation of the energy of falling water into electricity has improved markedly over the years to the point where efficiencies of 92 per cent have been achieved. Similarly, the process of transforming the mechanical energy of the turning steam turbine shaft into electricity in the generator approaches an efficiency of 98 per cent (reference 8, p. 95).

The problem that arises in heat engines, however, is considerably different in nature. The maximum over-all theoretical efficiency which can be achieved by even an ideal heat engine is much less than 100 per cent, and the maximum efficiency of a perfect steam engine is substantially less than that of an ideal heat engine. The models of these engines that can be economically constructed in the real world fall even further short of perfection, resulting in the seemingly low levels of efficiency prevailing today. To reiterate: the low efficiency of the steam cycle is

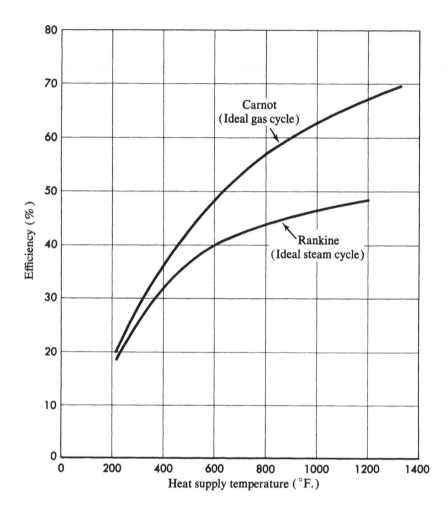

Figure A-1. Thermal efficiency of ideal heat engines. (Heat discard temperature taken as 76°F.)

due not only to those normal obstacles to perfection that can be at least partially overcome over the course of time, but also to other obstacles which, according to one of our most confidently held physical theories, are insurmountable. This is developed in detail below. Figure A-1 indicates the relative importance of these limitations.

Since these theoretical obstacles are so important in understanding the limits to improvements in heat efficiency, and hence in water use, we will attempt to explain them in this Appendix. But first, we will have to examine some concepts of temperature and heat.

TEMPERATURE AND HEAT ENERGY

In the course of time several temperature scales have been developed, each of which measures the heat intensity of a body relative to a certain standard. Thus, the knowledge that the temperature is 20° Centigrade means that this temperature is greater than that of ice at atmosphere pressure by 20/100 of the difference between the temperature of freezing water and that of boiling water. The Fahrenheit scale in the same fashion establishes a scale based on the coldest temperature Gabriel Fahrenheit could develop by freezing a salt water solution. In neither case, however, does 0° indicate an absence of all heat; the numbers on these temperature scales give a frame of reference, not an absolute measure. Furthermore, temperature measures only the intensity, not the quantity, of heat. A gas can be heated without raising its temperature or cooled without lowering it. Similarly, a rise in temperature need not necessarily imply the addition of heat. To illustrate by analogy, the relationship of temperature to heat is similar to the relationship of gas pressure to gas quantity, as a simple interchange of terminology in the above paradoxes will demonstrate. To understand these relationships more clearly, however, we turn to a short discussion of the concept of heat.

The physicist defines the thermal energy in a substance as the energy of its molecules, both in the form of the kinetic energy of motion or the potential energy of resistance to the molecular attraction of other molecules. In the case of gases, almost all of this energy is kinetic and is measured by the same mass-velocity relationships with which one would measure the energy of a batted baseball. In a solid substance much of the energy of the more closely packed molecules goes to offset the attraction of other molecules, that is, the energy is potential rather than kinetic. In both cases, to heat a substance increases its energy content. In the case of gases, this energy results in still greater movement, measured in part by the larger number of collisions of the gas molecules with neighboring objects, i.e., greater pressure. In the case of solids, the increased energy imparts the molecules with a greater ability to draw farther away from one another, permitting the readily observable expansion on heated solids. Further heating may cause the breakdown of molecular attraction to the point where the material flows, or melts. Still further heating may cause some molecules to break completely away from the surface of the material, evaporating or becoming a gas.

In the other direction, a substance can be said to have no thermal energy at all only when its molecules are completely motionless and offer no resistance to the attraction of their neighbors, thus becoming quite dense. This state occurs for all substances at the same minimum temperature, −459.7°F. (−273.2°C.). Because it is impossible, theoretically as well as practically, to find any temperature lower than this point it is known as absolute zero. Short of that temperature, heat, or more precisely thermal energy, remains in all materials. Considerable quantities of heat can be extracted from bodies at extremely low temperatures (say, −60°F.) by bringing them in contact with other substances at still lower temperatures. Two temperature scales have been set up on the basis of this absolute zero: one, the Kelvin, uses degrees of the size in the Centegrade scale and another, the Rankine, uses the Fahrenheit degrees. The freezing point of water is thus about 273 degrees Kelvin and 492 degrees Rankine.

We have noted that the motion of gases impinging upon their container constitutes what we call pressure. If we double the size of a container without increasing the number of molecules in it and without permitting a temperature change, the number of molecules that will collide with any particular portion of the container in a given interval will fall by one-half. Thus, while the quantity of gas remains the same, its pressure declines. Another way to reduce the pressure of the gas in this container is to remove heat from the gas by transfer through the container wall. As the temperature of the gas decreases, at constant total volume, the molecules move more slowly and there are fewer impacts on the walls in a given interval. Hence, the pressure decreases. All gases, with some generally minor modifications, obey the same rule with regard to pressure, temperature, and volume. This law is

$$PV = RT \qquad \text{(A-1)}$$

where P is pressure, V is volume, T temperature (measured in one of the scales based on absolute zero) and R is a constant dependent on the mass and type of gas.

IDEAL HEAT ENGINE CYCLES

Heat is merely one form of energy and is convertible into any other form of energy. The problem is to select the optimum technique for transforming heat energy into the mechanical variety. Assume that we have a heat supply at a maximum temperature F_1 and that we have a large heat receiver (such as the atmosphere or a river) at a low temperature F_2, without specifying the temperature scale for the moment. In order to use this heat at maximum efficiency, we want to supply it to a machine and then transform as much of that heat into work as possible before we exhaust the unusable heat at the lower temperature. In order that we may see the necessary relationships clearly on a graph, let us consider a variable, S, called "entropy," defined so that

$$TS = Q \text{ or } S = \frac{Q}{T} \qquad \text{(A-2)}$$

where T is temperature, and Q is the heat input at that temperature.[1] Thus, if we plot T against S on a graph, we can observe the heat inputs and outputs as *areas* under the curves. If the heat engine is to be self-contained and is to operate continuously, the cycle must be a closed one, i.e., whatever the behavior of the working fluid in the way of changes in pressure and temperature over the cycle, it must eventually return to its starting point. In order to find the useful energy output of our machine in the form of work, we must find the gross energy originally put into the engine as heat, less that energy which must be removed as heat to bring the fluid back to its original conditions after it has done its work. Such a complete cycle may be depicted on a graph as a closed polygon, the area of the polygon being the ideal net work output. The total heat supplied is represented by the area beneath the upper curves, down to the zero temperature axis. Figure A-2 gives several examples of such cycles. To find the optimum cycle, we must select the shape for

[1] More precisely and generally, $dS = \frac{dQ_{rev}}{T}$ where Q_{rev} is the heat added to the system in a reversible process. A reversible process is any ideal frictionless process taking place at conditions such that the slightest decrease in the potential causing it (such as the pressure of an expanding gas) or increase in the opposing potential (such as the load on the piston in the cylinder in which the expanding gas is confined) will cause a reversal in the direction of the process. The *change* in entropy between two conditions is thus

$$\int_1^2 dS = S_2 - S_1 = \int_1^2 \frac{dQ_{rev}}{T}.$$

If a reversible process takes place at constant temperature, such as the isothermal ideal expansion of a gas in a cylinder, $S_2 - S_1 = \frac{Q_{rev}}{T}$. Thus, the entropy of a gas, above its entropy at a reference condition at the same temperature is $\frac{Q_{rev}}{T}$.

which the enclosed area is the largest fraction of the total area between the upper curves and the zero temperature axis. It is readily seen that the maximum area ratio is that of a rectangle, as is shown in Figure A-3.

What are the implications of this perfect cycle? It tells us that the heat addition process, AB, must all take place at the highest temperature of the cycle and that whatever waste heat is ejected along the path CD must all leave the cycle at the lowest possible temperature. The other two parts of the cycle must take place at constant entropy, i.e., with no heat transfer to or from the surroundings. All four steps must be carried out in a hypothetical frictionless, reversible manner. This idealized process for converting heat to work is known as the Carnot cycle.

The Carnot Cycle. The optimum manner of converting heat to work is as follows: Take a working fluid at temperature T_1 (point D, Figure A-3) and compress it, thus increasing its temperature and pressure. The compression is carried out adiabatically and reversibly, i.e., without heat loss to the surroundings. When its temperature reaches T_2 (point A) stop compressing it and expose it to a heat source at temperature $T_2 + dT$. As the heat at temperature T_2 enters the fluid, permit it to expand so that the temperature remains constant, the working fluid doing work (e.g., lifting a weight) while expanding. Then (at point B) shut off the heat source and permit the gas to continue to expand, ideally, losing temperature and pressure, but without heat transfer with

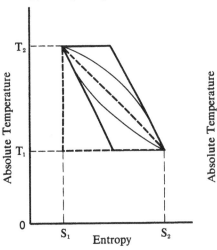

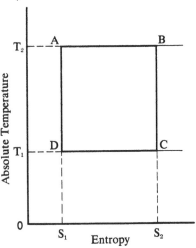

Figure A-2. Theoretical heat engine cycles.

Figure A-3. Ideal heat engine cycle (Carnot engine).

the surroundings. When the fluid has fallen to temperature, T_1 (point C), start to compress the fluid, permitting it to give off heat at temperature T_1 to some large heat receiver, such as cooling water, at $T_1 - dT$. Finally, at point D, stop this last process and begin again at the beginning. This most efficient heat engine is operating on what is known as the Carnot cycle. Its efficiency is the area $ABCD$ (work output) divided by the area ABS_2S_1 (the heat input). The heat discarded is the area DCS_2S_1.

If we follow these instructions accurately, we have performed the most effective transformation of heat into mechanical energy that is possible between the temperature limits T_1 and T_2. Nevertheless, this process will not yield a complete transformation of heat into work unless the exhausting temperature happens to be absolute zero. In order to see why this is so and to formulate an expression for the maximum efficiency, we must look more closely at the manner in which the work of the machine is accomplished. To measure the efficiency of the machine, we compare the heat energy supplied to the machine with the work output of the machine. Since the heat supplied to an ideal gas is exactly equal to the work done by the gas if the temperature of the gas is constant, the two isothermal heat transfer steps in the cycle may be viewed as work transfer steps also. Thus, the heat added to the gas at T_2 is equal to the work done by the gas in expanding along AB. Similarly, the heat removed from the gas at T_1 is equal to the work done on the gas by the machine along CD.

In both the cases of inputs and outputs, work can be measured as the product of a force and the distance through which it operates: in the heat engine, the force corresponds to the pressure of the gas and the distance is the change in volume, changes in both temperature and pressure being effected by the addition of heat. Since both pressure and volume are continually changing, the work can be best expressed graphically as the area under a curve of pressure vs. volume. Work is done by the gas on the machine and, in turn, on the surroundings as the gas expands. In order to compute the net work delivered per cycle we must properly subtract out the work that is required to return the gas to its original compressed state. If we were to add all the heat to the engine at the high temperature, the pressure-volume curve would be a hyperbola of form $PV = RT_2$. After all the heat had been added and we had permitted the gas to continue to expand and to do work, the temperature would reach the low temperature of the cycle. Now, in order to have a machine with 100 per cent efficiency, we would have to be able to return the fluid to its original pressure and volume without doing any work. That is, we would require that

$$PV = 0. \qquad \text{(A-3)}$$

But since R is a constant greater than zero, this condition will not exist unless $T_1 = 0$. The behavior of gases is such that this condition will not hold except where $t = -273°C$., so that no machine could achieve 100 per cent efficiency except with an exhaust at that extremely low temperature. But there is nothing so cold in nature to which heat would flow at this temperature and there is no possibility of achieving such a temperature without the expenditure of additional work, so the impossibility of 100 per cent efficiency becomes clear.

Efficiency of the Carnot Heat Engine. Returning to Figure A-3, we can now see how to calculate the maximum efficiency. The proper scale for the temperature has been established to be one of the Absolute Zero scales, either Kelvin (0°K. = -273°C.) or Rankine (0°R. = -460°F.). As a result we can measure the total heat input into the machine as the area ABS_2S_1, equal to $T_2(S_2 - S_1)$. The work done in the machine, i.e., the heat that is transformed into work, is the area $ABCD$, or $(T_2 - T_1)$ $(S_2 - S_1)$. The efficiency of an ideal, frictionless, reversible engine, the Carnot engine, therefore is

$$\frac{(T_2 - T_1)(S_2 - S_1)}{T_2(S_2 - S_1)} = \frac{T_2 - T_1}{T_2}. \qquad \text{(A-4)}$$

It can readily be seen from the formula that efficiency increases as T_2 increases and/or T_1 decreases, but that efficiency cannot reach 100 per cent unless T_1 were to equal zero on the Absolute Scale.

Unless we should locate all of our thermal-electric plants in the polar regions, we will find that it is impractical to achieve a year-round exhaust temperature much different from about 60°F., the average annual temperature of air or water. In terms of the Rankine scale, this temperature is roughly 520°, substantially different from zero, thus placing an effective limitation upon the efficiency of any heat engine. As an extreme upper limit for any heat engine using ordinary fuels, we may take the combustion temperature of carbon fuels at say 3,000° Rankine. Between these limits, the maximum theoretical efficiency that can be attained is about 83 per cent, no matter how well constructed the apparatus. If we take as the upper limit the 2,000° level which constitutes the maximum temperature which can be contained for any length of time by existing metals, we find that the maximum efficiency drops to around 74 per cent. With the highest temperatures in use in steam cycles (1,660°R.), the efficiency of the ideal Carnot cycle is only a little over 68 per cent. It will be further noticed that with any given exhaust temperature, constant increases in the initial temperature yield successively smaller increases in the thermal efficiency of the Carnot cycle.

Against this last standard, the maximum efficiency so far attained in a steam cycle, that is, around 42 per cent, looks considerably better than it does at first glance. In other words, the most efficient steam cycle yet attained is about two-thirds as efficient as that of the ideal heat engine operating within the present limits of temperature. The performance of modern steam stations is even more striking if we recall our earlier statement that the Carnot efficiencies are not the most strict limitation on the possibilities for reducing thermal losses in a steam station. The use of steam as a working fluid in our heat engine reduces the maximum attainable efficiency still further and raises, by comparison, the current level of achievement. We will now turn our attention to a discussion of the limitations on efficiency that are introduced by the use of steam as a working fluid and the reasons why steam is nevertheless used in these machines.

The Rankine Cycle. We have seen that for maximum efficiency in a heat engine, all heat has to be added to the working fluid at the highest temperature and emitted at the lowest, and each of the four steps in the process must be conducted so that the maximum work output (and minimum work input) is obtained. In order to accomplish this, the heat engine must be designed so that during heat transfer to the fluid, expansion against a moving piston or some other load or force must take place. Then the gas must continue to expand against the piston, but with no heat transfer occurring. Similarly, the heat rejection process must be accomplished simultaneously with compression of the gas, and the final compression to the starting point must be done with no heat loss. Moreover, for a perfect Carnot cycle to be hypothetically followed, the so-called perfect gas is required, for only with such a fluid is the work done equal to the heat transferred in the two isothermal steps of the process.

Although the above ideal Carnot cycle can be fairly well approximated by the so-called hot air engine (also known as the Stirling engine and the Ericsson engine), this machine was bulky, expensive, and used for only a limited time in the nineteenth century. The design problems associated with the direct heating and cooling of air confined in the cylinder of the machine were the principal factors leading to its replacement by the steam engine.

Steam is not a perfect gas. It evaporates and condenses well within the range of temperature that are used in heat engines. As a result of this, in at least part of the ideal steam cycle (the low-temperature portion), the fluid used is in the form of water—a practically non-compressible liquid which does not conform to the Carnot requirements. This is both an advantage and disadvantage. While undergoing vaporization and condensation, heat transfer can take place at constant temperature even though no work is being done. Thus, these steps need not take place in the engine itself, but rather in simple boiler and condenser units. However, as a result of liquid being present in parts of the cycle, the ideal efficiency of this process is below that of the perfect Carnot cycle.

The ideal steam cycle is known as the Rankine cycle. The fluid (water) enters the cycle at the lower temperature, T_1, shown in Figure A-4 as point A. The dashed curve is the boundary of liquid and vapor phases: to the left is compressed liquid, within the loop is liquid and vapor mixture, and to the right is superheated steam. Thus point A represents liquid water discharged from the condenser. This liquid then enters the boiler feed pump where its pressure is raised sufficiently for it to flow into the high pressure boiler, at point B. The water temperature remains nearly at the condenser discharge temperature during this step. After the water enters the boiler at point B, its temperature increases along the path BC as it is heated by fuel. When it reaches the temperature corresponding to point C, (that is, T_2) it starts to boil, and as it is converted to steam at constant temperature and pressure, it follows the path CD. At point D, the water has been completely vaporized. At this point, the steam enters an engine (turbine or reciprocating) and expands against the load, ideally without friction or heat losses, along the vertical path DE, to the low pressure and temperature corresponding to vapor and liquid in equilibrium with each other at point E. This condition represents a mixture of low pressure steam with a moderate amount of condensed water. At point E, this mixture enters the condenser where heat is removed, causing the steam to condense to liquid water, at constant temperature and pressure, until only condensate remains at point A. The condensed steam then becomes the intake water for renewing the cycle.

The major difference between this cycle and the ideal Carnot cycle is that a portion of the heat must be added to the fluid while it is changing

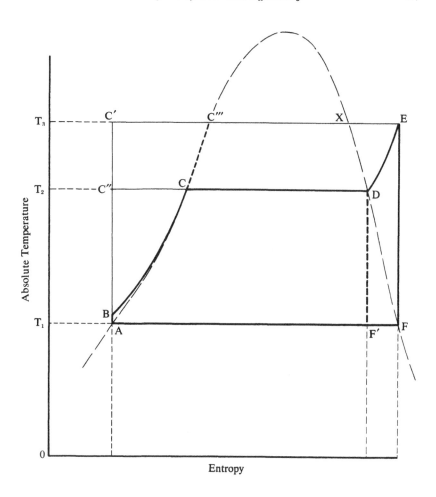

Figure A-4. Ideal steam engine cycle (Rankine engine).

in temperature, (along path *BC*) thus making it impossible to extract the maximum (Carnot) work out of the cycle. In other respects, such as the constant temperature heat supply to convert water to steam, the expansion to low temperature with no heat loss, and the constant temperature heat rejection in the condenser as steam is converted to water, the process follows the Carnot ideal exactly. As we shall see later, it is not possible to achieve these ideal conditions in practice, but these problems are of the ordinary mechanical and engineering type which restrain all attempts to produce perfect efficiency and do not present us with a theoretical obstacle to high efficiencies.

There is also a minor difference in the two cycles. The compression of liquid water along path AB involves very little energy input and causes very small changes in all properties and conditions, except the pressure of the water. The high pressure corresponding to the upper conditions at T_2 is reached in this liquid pumping operation with very little work expenditure. In the Carnot cycle, however, a large quantity of energy is supplied in this step (path DA in Figure A-3). But it is not this work input difference which is the significant item, it is the difference in the constant-temperature portion of the heat supply step, CD versus $C'D$ in Figure A-4.

It would be theoretically possible to operate a Carnot cycle with a water-steam working fluid, by following the path $A'CDE$. This would require halting the condensation step EA' when only a portion of the low pressure vapor had been condensed, and then compressing (with no heat loss and with "left-over" work from the expansion step DE) this mixture to the pressure and temperature of point C. Constant temperature heat supply and subsequent expansion to E would complete the cycle. Although this cycle would have higher efficiency than the Rankine cycle, it is mechanically difficult because of partial condensation requirements and because the compression step $A'C$ would require large and expensive auxiliary compressors.

Efficiency of the Rankine Heat Engine. The efficiency of the Rankine cycle, like any other cycle, is equal to the net work delivery divided by the total heat supplied. Referring to the T-S diagram, it is therefore the work delivered in the expansion step DE minus the work supplied to the liquid water along AB, this difference divided by the heat supplied in the boiler for water heating and steam generation, along path BCD. For practical purposes, the work input in step AB may be considered negligible, so the difference in energy content of the fluid from D to E divided by the difference from A to C is equivalent to the efficiency. Graphically, the efficiency of the Rankine cycle is the enclosed area $ABCDE$ divided by the area $ABCDEFG$, equal to the work output and heat input respectively. If the Carnot cycle were followed, the efficiency would be the area $A'CDE$ divided by $A'CDEFG'$. Comparison of these area ratios will show the Carnot to be the larger of the two.

Anyone who has ever boiled water, however, has undoubtedly noticed the length of time required for a substantial amount of water to boil away after the water has reached the boiling temperature. The heat that is required to shake the water molecules loose from the water after it has reached the boiling temperature is called the latent heat of evaporation.[2] At normal atmospheric pressure, this latent heat far exceeds that required to bring the water to the boiling temperature. For example, starting from water at 50°F., the heat required to bring the water to 212° is only (212 − 50) or 162 Btu, while the heat required to vaporize that 212° water is 970 Btu, or almost six times as much. This massive addition of heat at the boiling temperature is, as a result, an important factor in approximating the Carnot cycle with a steam fluid.[3]

The heat of vaporization does not remain constant for all steam conditions. As pressure increases, the temperatures required to vaporize the water also increases. At higher temperature and pressure the energy content of saturated steam is greater, but the heat that is added to the water before it vaporizes increases even more rapidly, so less heat is required for vaporization at the higher temperature and pressure. Since it is this large addition of heat at a constant temperature which permits the steam cycle to approximate the ideal, this decline in the latent heat as temperatures increase means that the two cycles diverge more and more at higher and higher temperatures.

This process can be traced in Figure A-5. The vertical scale gives the temperature in degrees Fahrenheit; the horizontal scale is the energy content of the fluid (by convention liquid water at 32°F. is taken as zero). The figure formed by the pressure lines outlines the different phases through which water passes in the heating process. Along line NP water is always in its liquid phase; in the area to the right of line PB it is always in the form of steam; inside the loop it is a mixture of both, i.e., it is boiling or condensing. (The area above the point P will be discussed later.) The dotted lines running through the diagram illustrate the behavior of water at various pressures.

Take, for example, one pound of water at 32°F. in the bottom of a large, heavy-walled cylinder in which a loaded piston exerts a total absolute pressure of 500 pounds per square inch (psi) on the water. No air or other material is present, and no water vapor can exist at this temperature and pressure. If we add heat to the water, we would follow

[2] Actually some of the molecules will have already shaken loose at still lower temperatures, but these will represent only a small fraction of the total.

[3] It should be noted that while pressure does not enter explicitly into our graph, boiling water at T_2 implies a certain pressure which automatically exists as long as boiling is taking place.

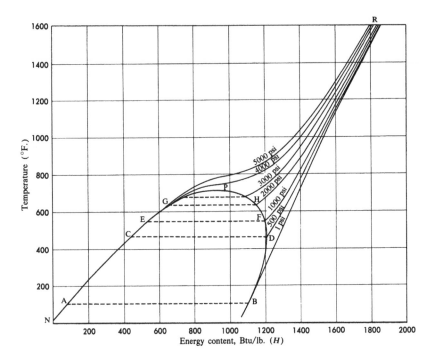

Figure A-5. Temperature-energy changes in Rankine cycles.

the line *NC* which, as we can see, is practically a straight line in this range. The slope of this line, the reciprocal of the specific heat of water, indicates the rate at which heat addition increases the temperature. When the temperature reaches 467°F. (point *C*), the vapor pressure of the water will have reached 500 psi, the same as the imposed pressure, and further additions of heat will cause the water to boil, without additional temperature or pressure change. Inspection of the horizontal scale shows that 449 Btu will have been added to the water during this heating process. (Reference 37.)

As steam is produced on further heat supply, the piston will be pushed upward until all the water is vaporized. This process, along the constant temperature constant pressure path *CD*, will require the addition of 755 Btu additional heat, to a total of 1,204 at point *D*, where one pound of saturated steam at 467°F. and 500 psi exists. Further additions of heat, beyond this point will increase the temperature once more, producing superheated steam[4] but, as we can see from the line DR, its slope (the reciprocal of specific heat of *steam*) is steeper and not quite so constant as the specific heat of water. The line CD, representing 755 Btu, is the latent heat at that temperature. In the same fashion, AB, EF, and GH represent the latent heat of vaporization at 1 psi, 1,000 psi, and 2,000 psi, and are equal to 1,036, 649, and 463 Btu respectively.

As the condition of the saturated steam approaches 3,206 psi and 706°F., the latent heat becomes very small, until finally the latent heat disappears. What this means is that the addition of heat to water at or above those conditions results in a continually rising temperature without any of the phenomena which we associate with "boiling." There is no noticeable physical change in the substance as one passes from a lower temperature to a higher one. This process may be visualized in Figure A-5 by considering the heating of liquid water at a constant pressure of 5,000 psi. We would thus follow nearly along (very slightly to the left of) the water curve $NACE$ and then move away from it along the 5,000 psi curve toward the region indicated by R. There was no horizontal portion in this path, hence no constant temperature boiling. Although this concept of non-boiling water may strain our imagination a little, it causes no difficulty in a physical form, with the resulting fluid behaving almost exactly like steam in the machinery and with determinate physical properties. For our purposes it is important now only to point out that the phenomenon of latent heat eventually disappears as the temperatures and pressures rise.[5]

There is, of course, no rule which requires us to add all of the heat of vaporization at the upper temperature of the cycle. It would be possible to construct a cycle such as $ABCDEF$ in Figure A-6 which continued to heat the fluid after it had turned completely to steam. Now it is clear that the simple Rankine cycle $ABCDF'$ comes closer to the Carnot ideal $AC''DF'$ than the higher temperature cycle $ABCDEF$ comes to its Carnot twin $AC'EF$. But since the Carnot cycle that operates between T_3 and T_1 is more efficient than one which operates between T_2 and T_1,

[4] Superheated steam may be considered steam which has been heated to a higher temperature than that at which it was boiled from liquid water, but at essentially the same pressure.

[5] The concept of a steam beyond the critical temperature is not an idle theoretical point. Several large power plants have such supercritical cycles either completed or under construction. It is also worthwhile to note that while the supercritical fluid does behave much like lower temperature steam, some unique and difficult engineering problems exist, some of which are important only in the immediate neighborhood of the critical point.

it is still possible that the cycle using the hotter steam, *ABCDEF*, will be more efficient than the one described in Figure A-4 and shown in Figure A-6 as *ABCDF'*. Of course, from a theoretical point of view it would be still better to use a third cycle (*ABC'''EF*) than either of the others. However, if it costs money or resources to increase pressure (point *C'''* corresponding to higher pressure than *C*) it might be eco-

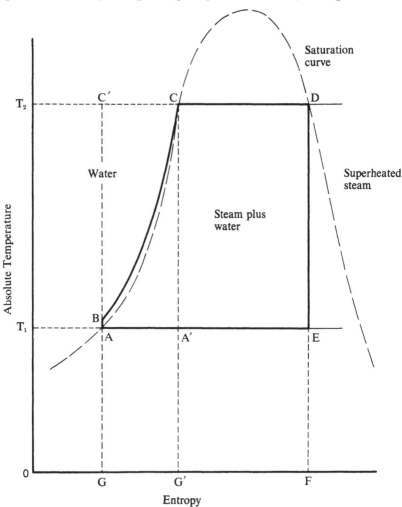

Figure A-6. Comparison of Rankine, Carnot, and hypothetical heat engine cycles.

nomically more efficient to use a cycle with steam raised above its boiling point (point E being at the same pressure as points C and D). Moreover, the physical accomplishment of isothermal steam expansion in an engine, from X to E, while heat is being added to the steam, would be practically impossible.

Returning now to the ideal Rankine cycle depicted in Figure A-4, the line DE represents the ideal expansion of the steam against a load, which lowers its temperature without giving up heat to or receiving heat from the external environment. The line EA reflects the removal of heat from the fluid at constant temperature, which, in the Rankine cycle coincides with the condensation of the steam into water.

A COMPARISON OF THE CARNOT AND RANKINE CYCLES

Since the last two stages of the Rankine and Carnot cycles are so similar, comparison of their relative efficiencies is simplified. In the Carnot case, we found that because both the areas of heat addition and of heat removal were rectangles with common width, we could compare the two areas by referring only to the ordinates; i.e., the temperatures. In the Rankine case this is not possible unless we substitute for the irregular polygon, the rectangle to which it is equivalent. This can be approximated by taking as the ordinate of the hypothetical rectangle the weighted average temperature with the weights being the heat added at each temperature. The advantage of this apparently cumbersome process is that once attained, this average temperature can be used in the Carnot formula and will facilitate the interpretation of the factors affecting the efficiency of the Rankine cycle.

Let us note the effect of latent heat on the efficiency of low-temperature steam units. Theoretically, a steam engine operating at a top temperature of 212° and an exhaust temperature of 80° would add 132 Btu at an average temperature of $(212 + 80)/2$ or 146°F., and would add 970 Btu at a temperature of 212°, thus making a weighted mean temperature of heat addition of 203°F. or 663°R. The Rankine efficiency of this cycle, then, is:

$$\frac{663 - 540}{663} \quad \text{or} \quad 18.6\%^{[6]}$$

---

[6] This computation is not completely precise because of a slightly varying heat capacity of liquid water as temperature changes. Exact determination can be made by use of tables or charts of the energy content of steam at various pressures and temperatures (reference 37).

while the maximum possible (Carnot) efficiency was only:

$$\frac{672 - 540}{672} \quad \text{or} \quad 19.7\%.$$

Thus, because of the substantial addition of heat at the upper temperature, the effective average temperature of addition is very close to the maximum and the Carnot efficiency can be closely approached by the Rankine.

This close approach to maximum efficiency was made possible only by use of a very low temperature cycle. As a result, we have theoretically achieved over 94 per cent of the *possible* efficiency only by using a maximum temperature which restricts that possible efficiency to less than 20 per cent. If we would raise the top temperature to 400°F. and adjust the pressure (to 247 psi) so that vaporization takes place at the maximum temperature, the Rankine cycle would drop to about 88 per cent as efficient as the Carnot cycle. Nevertheless, because the maximum (Carnot) efficiency increases from 19.7 to 37.3 per cent, this latter cycle is much more efficient (33 per cent) than either of the cycles at the lower temperatures. As we have indicated earlier, the increasing divergence between the Carnot and Rankine cycles is due to the diminishing value of the latent heat addition at the upper temperature, but this divergence is not enough to offset the advantages of moving to higher and higher temperatures so long as we are concerned solely with achieving maximum physical efficiency. It does, however, place an effective limit on the efficiency of a steam cycle which is considerably lower than the Carnot ideal and far short of 100 per cent. And when there are economic costs involved in increasing temperatures and pressures, the diminishing relative efficiency of the steam cycle places important limits on the economically feasible fuel and water savings.

EFFECTS OF PRESSURE AND TEMPERATURE ON RANKINE CYCLE EFFICIENCY

For any given maximum temperature the most efficient Rankine cycle is that in which the pressure is great enough to produce steam at the maximum temperature. For any pressure lower than that, the latent heat would be added (i.e., the water would boil) at a lower temperature and the average temperature of heat addition would be lowered. On the other hand, it should also be clear that the advantage of this cycle over one which uses a lower pressure and superheated steam declines as the temperature rises. The reasons are twofold. First, the latent heat of vaporization of water to form saturated steam declines rapidly as temperature rises; second, the specific heat of water being higher than that of steam, the fact that a greater portion of the heat is being used to raise the water temperature tends to encourage the heating of the fluid in the form of steam rather than in the form of water. Still, both of these factors only reduce the disadvantage of using superheat versus higher pressure, and if the cost of increasing pressure were zero, the use of saturated steam would be preferred (in an ideal cycle). It is easy to see, however, the possibility in the real world of accepting somewhat lower efficiency in return for lower expenditure for pressure increases, especially as temperatures increase.

Above the critical temperature there is, by definition, no latent heat. Nevertheless the phenomenon leaves its shadow in the form of some peculiarities in the behavior of the specific heat. At temperatures under about 500°F. the specific heat of water rises very slightly as temperature increases to the saturation temperature. After it reaches that temperature it suddenly becomes, in effect, infinite as the latent heat is added with no temperature change. Finally, as the water is all converted to steam, it changes back, once again discontinuously, to a level about half that of liquid water. As saturation temperatures rise beyond 500 degrees, there is an increasing tendency for the specific heat to rise as the water temperature approaches the vaporization temperature, thus reducing the discontinuity of the jump when we reach the temperature at which vaporization takes place.

Similarly, at high temperature and pressure, the specific heat of steam also is large, decreasing to its normally low value of 0.5 to 0.6 at very high temperatures (1,400 degrees or more). In the vicinity of the boiling point at high pressure and temperature, i.e., near the critical point, specific heats of both water and steam are high, and the discontinuity in this property in boiling water at these conditions is not great. The complete disappearance of the phenomenon of latent heat at the critical conditions is therefore indicated by a rapid rise in the specific heat of water in the lower neighborhood of 706° to a momentary value of infinity at that temperature, followed by a steep decline to lower values at much higher temperature. Above the critical pressure this sharp rise in the specific heat as temperature rises continues to occur, though with diminished severity, until for very high pressures the specific heat shows only a slow rise and fall over the whole range of temperatures. As a result of this, the theoretical advantage of raising boiler pressure at a given

temperature above the critical steadily decreases, until at some point
well beyond the range of the highest pressures currently in use it disappears entirely.

The behavior of efficiency with changes in the temperature and pressure conditions for the simple Rankine cycle is shown in Figure A-7.
Temperature in degrees Fahrenheit is on the horizontal scale and pressure in pounds per square inch on the vertical. Condensation at one inch
of mercury pressure, corresponding to a temperature of 79°F., is assumed. The curved lines on the graph are iso-efficiency curves: each curve
is the locus of all combinations of pressure and temperature which will
produce the same thermal efficiency. The pressure scale has been taken
as logarithmic because of the much wider range of pressure than of
temperature, and as a result some portions of the curves are concave
to the origin. The boundary line on the left of the graph shows the minimum temperatures (and corresponding pressures) at which the indicated
Rankine efficiencies are possible. As we have pointed out before, we

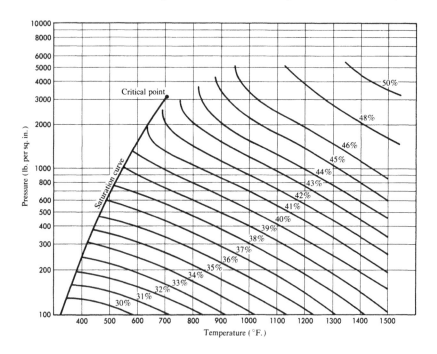

Figure A-7. Effects of temperature and pressure on the efficiency of a Rankine
engine.

would expect this line also to be the locus of saturation temperatures and pressures, and examination of the graph indicates that this is so. In the diagram we have not extended the curves to the left of the line, because in that region water cannot boil and a steam cycle is impossible. Above 706°F. there is no boiling point, and in that region the line becomes determinable only by calculation.

It may be observed in Figure A-7 that as the pressure is increased at constant temperature, efficiency rises, i.e., heat rate declines, and at temperatures below the critical temperature, 706°F., the efficiency reaches a maximum at the saturation pressure. At constant temperatures above the critical, efficiency rises continuously with pressure, but at progressively lower rates until there is very little increase even in a 1,000 psi increase from 4,000 to 5,000 psi. An increase in temperature at constant pressure may be followed by moving horizontally on Figure A-7. It is seen that in all pressure ranges, a rise in temperature results in an increase in efficiency which is reasonably uniform over all temperature and pressure conditions except very near the critical. Roughly, a 100°F. rise in temperature results in about a 1 per cent efficiency increase.

THE IMPORTANCE OF THE HEAT EXHAUST

So far, we have dealt almost exclusively with the effect of high initial temperatures on the efficiency of the conversion process, and in view of the historical pattern of improvement in cycle efficiencies there is a good deal of logic to this emphasis. Nevertheless, it is important to remember that the lower temperature limit, while less subject to human manipulation, is still a factor of considerable importance in determining the efficiency of ideal steam plants. Furthermore, in view of the importance of water in establishing the temperature of heat exhaust in current steam practice, it is even more important for us to fully understand the role played by the latter in the technology of electricity production.

The earlier diagrams have shown that the heat exhaust portion of the Rankine cycle corresponds closely to the constant temperature heat exhaust required by the Carnot cycle. These two portions are not identical, however. In the Carnot, pressure increases and volume falls as the heat is exhausted. In the steam cycle, the heat loss occurs at constant pressure, the volume of the steam falling more rapidly as the steam condenses to water. Despite this difference, the exhaust temperature affects the Rankine cycle in much the same way it affects the Carnot. If we replace the T'_2 of the Carnot formula by the average temperature of the heat addition process, T_A, the efficiency of the Rankine cycle can be stated

$$\frac{T_A - T_1}{T_A} \quad \text{or} \quad 1 - \frac{T_1}{T_A} \qquad \text{(A-5)}$$

so that decreases in T_1 yield increasing efficiency. If we look more closely at the formula we can easily see that the effect of any given decrease in T_1 is less the greater is the difference between T_A and T_1, so that increasing thermal efficiency by raising initial temperatures reduces the value of any given absolute decrease in the exhaust temperature.

Despite decreases in the value of temperature reductions at the exhaust end of the cycle, the savings are still important enough to warrant extensive capital investment to keep the exhaust as cool as possible. Obviously, the simplest way of disposing of the steam after it has done its work is to let it exhaust into the air. This would permit the steam to expand until the pressure fell to 14.7 psi, leaving the temperature of the exhaust steam at 212°F. This, in fact, is the way railroad steam engines operate. An alternative technique would be to condense the steam to water in a water-cooled condenser, reducing the condenser pressure to, say, 0.5 psi and the steam temperatures to 79°. If the upper conditions of the cycle were 115 psi, 338°F., the first technique would yield an efficiency of 13.1 per cent, the second would give us an efficiency almost twice as great (29.3 per cent). If the initial conditions had been, say, 1,000 psi, 545°F., the difference would have been 25.1 vs 38.6 per cent. At still higher temperatures, the difference falls still further, but at no point within the range of reasonable temperature does the difference become so small that condensation would not become advantageous unless cooling water were *very* expensive (as, for example, in the railroad steam engine).

METHODS FOR INCREASING THE EFFICIENCY OF IDEAL STEAM CYCLES

Feedwater Heating. Up to this point, we have discussed an ideal steam cycle of the simple type: one in which the Carnot procedure is followed as closely as is possible with steam. We have seen that this simple steam cycle falls short of the Carnot in efficiency. In this section we shall see how this cycle can be improved upon by recovery of some of the waste heat that might otherwise be rejected to the condenser.

In the simple Rankine cycle, the cold condensate receives heat from burning fuel, as in the boiler itself. It is clear that a heat source at a much lower temperature could be used for heating, partially at least, this boiler feed water. By so doing, less high-temperature heat could be needed and efficiency might improve. If an outside supply of steam at moderate temperature is used for this purpose, it is obvious that the cycle efficiency would be increased. As a matter of fact, if a *portion* of the steam in the cycle itself is first expanded to an intermediate pressure and temperature, doing useful work in the process, and then this steam is used for heating the condensate resulting from the balance of the steam expanding *all the way* to condenser pressure, the efficiency of this combination is greater than that of the simple Rankine Cycle. This is all possible because in any steam cycle heat is added over a wide range of temperatures. Thus, high pressure steam can be permitted to expand partially, doing some work; then, by using the remaining energy in the steam to add heat to the feedwater (there is no waste in this process), the average temperature of heat addition can be raised and efficiency increased.

The argument can be illustrated as follows. Say that water comes from the condenser at a temperature of 80°F. and is heated (by fuel) to 338°F. at a pressure of 115 psi. Some heat is added at every temperature between 80° and 338° at roughly the rate of 1 Btu per degree, so that the average temperature of heat addition is considerably less than the 338° maximum (209°F.). The maximum efficiency of this steam cycle in producing work is only 29 per cent. In another process, let 1 lb. of the 338° steam expand in an engine only part way, say to 212°. By that time it will have transformed into work about 15 per cent of the energy it received from the fuel. If we were to continue to expand it, another 14 per cent could be converted to work. On the other hand, the energy remaining in 1 lb. of 212° steam could be used to heat over 6.4 pounds of feedwater from 80° up to 212°F. By raising the feedwater temperature to 212°F., the average temperature of fuel-heat addition is raised to 330°F. since the heat addition from 80 to 212° has already been accomplished. This implies about a 31.8 per cent efficient cycle. Offsetting this, however, is the fact that a portion of the steam is not permitted to expand fully. Nevertheless, the gains offset the losses, and this so-called regenerative cycle offers an efficiency of 30.7 per cent as opposed to 29 per cent for the simple cycle.

Since feedwater heating improves the cycle only by making it approach the Carnot more closely, it is clear that feedwater heating cannot raise the Rankine efficiency above the Carnot. Actually, the maximum efficiency would have to be somewhat lower, since feedwater heating cannot exactly duplicate the Carnot and there is an offsetting loss from the unexpanded steam. Thus, the greater is the gain on the heat input side (i.e., the higher that the average temperature of heat input is raised), the greater is the loss in incomplete steam expansion on the output side, so that the marginal net returns from feedwater heating decline as more such heating is done. In theory, the Carnot efficiency could be equalled if instead of *one* stage of feedwater heating by partially expande steam, an infinite number of stages were used, each differing from the next only by an infinitesimal temperature. In practice, particularly in modern high-pressure and -temperature cycles, several stages of feedwater heating actually are used, employing partially expanded steam extracted from the turbines at various pressures. The exact formulation of the ideal regenerative cycle can be developed mathematically, but for the purposes of this study, this explanation must suffice.

Superheating. For any specified operating temperature, the most efficient steam cycle would be that which operated at the saturation pressure. This, however, is not an optimum operation unless increases in pressure are costless, and the true costs of increasing pressure and temperature are such to make operation at saturation pressures uneconomic except at very low pressures. Since, for any given pressure, there is no similar limit on the extent to which temperature increases may increase efficiency, the practical situation is typically one in which the temperatures used are considerably above those at which the steam would be saturated. As a matter of fact, the use of temperatures higher than the saturation temperature (i.e., the use of superheated steam) is due not only to the relative costs of increasing temperature and pressure but to some technical inefficiencies in the extraction of work from saturated steam as well. These mechanical inefficiencies are not unimportant and later will be discussed more fully. Here, however, the major point to make is this: that economic considerations may lead to the use of superheated steam cycles and thus create the possibility of further refinements in the simple cycle.

The use of superheat may be conveniently visualized in Figure A-8 representing the simple Rankine cycle with the addition of the superheating step. If saturated steam is delivered from the boiler (point *A* on the diagram) to a superheater consisting of banks of tubes heated by the hot furnace gases, the temperature of the steam is increased at constant pressure along path *AB*. This heating process results in the addi-

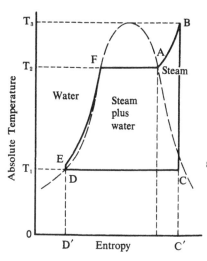

Figure A-8. Ideal steam cycle with superheating.

tion of energy to the steam, an addition at an advantageously high temperature. Thus, the superheated steam contains *more* energy than saturated steam, and the average *temperature* of energy absorption by this steam is higher.

Next, the superheated steam is supplied to the turbine in which it is expanded (ideally) along path BC, delivering work to the turbine shaft, and finally passing to the condenser at the low pressure and temperature T_1 obtainable by cooling water use. The balance of the cycle along paths CD, DE, EF, and FA is identical with the simple Rankine cycle.

As with the use of saturated steam, the efficiency of the superheat cycle is very nearly equal to the work delivered in the expansion step, BC, divided by the total heat supplied to the steam, which is the sum of water heating, EF, steam generation, FA, and superheating, AB. Thus, the decrease in the steam's energy content B to C divided by the increase E to B is the efficiency. Graphically, this is equal to the enclosed area $ABCDEFA$ divided by the area $ABCC'D'EFA$. It may readily be seen that the portions of these two areas under the superheating step AB are such that this part of the energy received by the steam is being used at higher efficiency than the balance. The result is a higher efficiency for the superheat cycle than for the Rankine cycle operating at the same maximum pressure. Since the cost of superheating is not excessive in comparison with the gain in efficiency, practically all power plants follow this practice.

Resuperheating or Reheating. Just as superheating permitted an improvement in thermal efficiency without resorting to higher pressure operation, a second and third superheating of steam after partial expansions through the turbine may be used to achieve still higher efficiencies.

The technique that makes this improvement possible can be described as follows. Take the superheated steam and start to expand it so that it performs work. After it has partially expanded, return the steam to a heat source and raise its temperature again, this time at the lower pressure. Having raised its temperature, now permit it to undergo further expansion in another turbine operating at lower inlet pressure (or in the low pressure stages of the same turbine), and exhausting into the condenser. The process we have described is *single* reheat—there is no theoretical limitation on the number of times the steam can be reheated. On the other hand, while this process will improve the efficiency of steam cycles, there is no automatic guarantee that it will do so. In most high-pressure steam plants, the intermediate pressure or pressures must be carefully chosen or else slight decreases in efficiency can result. The net objective being sought is the supply of additional heat at the highest possible temperature in order to raise the effective average temperature of the heat supply.

The improvements in cycle efficiency come, in effect, from adding to the original cycle another cycle with a higher efficiency (see Figure A-9). Since this latter cycle is similar in many respects to the Rankine cycle, we can calculate its efficiency in much the same fashion. In the diagram the steam is first expanded down to approximately the saturation pressure and temperature which it would reach if no heat were added or subtracted in the expansion process. This steam is then reheated to the original temperature.[7] It can be seen that the average temperature of the heat addition will be considerably higher than if the steam had been condensed and returned to its original pressure: that is, the efficiency of work production by the reheat process will be greater than if we produced the extra output by expanding the steam fully and repeating the process. We have added a cycle of greater efficiency than the original and paid

---

[7] There is no physical reason why the reheat temperature must be the same as the original temperature, although in most cases it is the same. In practice, the same limitations that accounted for the limit of the initial temperature operate to limit reheat temperature, while the higher the reheat temperature the greater the efficiency. Because the reheated steam is at a lower pressure and hence occupies a greater volume, there are sometimes enough capital costs involved to keep the reheat temperature slightly below the initial level.

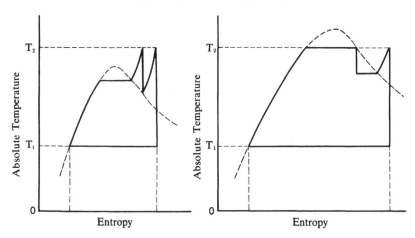

Figure A-9. Ideal steam cycle with resuperheating.

Figure A-10. Effect of resuperheating in a saturated steam cycle.

for it by the work foregone in not letting the steam expand fully from the initial high pressure.

In looking at the diagram, we can deduce two further characteristics of reheat. First, notice what would have happened if we would have permitted the original steam to expand to T_1. The steam would then have fallen below the saturation temperature and pressure. As a result, the reheat process would have included a substantial addition of latent heat at the lower temperature before the temperature would have started to rise. From what we know about the efficiency of heat cycles, we can see that this would have substantially lowered the efficiency of the reheat cycle. This is particularly important because since the reheat pressure is lower than the original pressure, the temperature at which water boils is also lower, thus making it possible that the reheat cycle may be even less efficient than the original if the reheat starts with appreciable liquid water at saturation pressure and temperature. This possibility is indicated more clearly in Figure A-10. If saturated steam is expanded to an intermediate pressure (short vertical line in Figure A-10) it partially condenses, becoming "wet steam." Withdrawal and reheating at constant pressure results first in constant temperature vaporization of the moisture (short horizontal path to the saturation curve) and only then can the temperature rise back to the initial temperature. The additional work obtained in the final expansion will seldom represent as large a proportion of the additional heat addition as if no reheating had been employed. As a corollary to our initial proposition, we can therefore see that reheat is of little use except in combination with superheat. Thus, while reheating can bring superheat cycles more closely in line with Carnot performance, it could seldom improve saturated steam cycles, even if we ignore economic factors.

REAL HEAT CYCLES

The discussion to this point has dealt with the ideal types of heat cycles which can be utilized to produce work. Even though actual steam cycles typically fall short of the ideal by not inconsiderable amounts, the concept of an ideal cycle is useful in delimiting our ideas of the range of possible improvement in the thermal efficiency of electric power generation in the years to come. Now that we have established that wider range, we can look more closely at some of the considerations tending to reduce still further the possibilities for cycle improvement.

In dealing with the ideal cycles, we have neglected energy losses in the important pieces of machinery which are used to transform the heat energy into useful work. In the most common heat cycle used to produce electric power, heat addition takes place in a steam boiler and the output is changed into mechanical energy in a turbine and into electrical work in a generator. The steam exhausts into a condenser and the condensate must then be pumped into the boiler. Each of these units is subject to some inefficiencies which will be discussed in the next several sections.

Boiler Feed Pumps. In the discussion of the ideal Rankine cycle, it was pointed out that because of the incompressibility of liquid water very little energy is theoretically required to pump the condensate back to the high pressure boiler. While this work is small relative to that produced by expansion of the steam, it is nevertheless a draft upon the output of the heat engine. The addition of the feed pump work requirement on the energy input side means that the true Rankine efficiency is slightly lower than that computed by considering only the steam generation and expansion steps. For example, raising the pressure of one pound of liquid water at 85°F. from atmospheric to 1,000 psi would theoretically require only about 3 Btu of work, which would be about 1 per cent of the work obtained from the steam on expansion in an efficient engine.

In modern steam plants, where several feedwater heating stages are used, boiler feed pumps handle water at relatively high temperatures and pressures. As the water temperature rises, its density decreases. Since the work required to raise the pressure of a fluid increases with an increase in pressure and a decrease in density (i.e., an increase in volume), the work done by the pump rises substantially at the higher temperatures and pressures. The ideal work requirement for pumping water at 500°F. to 3,000 psi, for example, is about 9 Btu per pound or 2 to 2.5 per cent of the output.

In addition to the theoretical energy requirement for boiler feed pumping is the work for overcoming inefficiencies in practical pumping equipment. At the severe conditions of modern steam power plants, actual feed pump power consumption can be considerably higher than the theoretical. Careful design and maintenance are therefore justified in reducing these losses and maximizing over-all efficiency.

Turbine Efficiency. In order to understand the sources of efficiency losses in a turbine we must first understand a few of the basic principles of turbine operation. The expansion of the high-pressure, high-temperature steam involves the motion of a gas with a certain mass through a distance. The principle of the turbine involves an attempt to transform the internal energy of the hot steam into the rotary motion of the turbine shaft. The method by which this transformation takes place is analogous to the process by which windmills operate.

In the case of the windmill, the pressure of the wind pushes against inclined blades, causing them to turn. In a similar fashion the high pressure steam from the boiler is permitted to expand while pushing against similarly inclined blades inside the turbine. The turbine blades, in turn, are all connected to a central shaft which rotates under the pressure on the blading.[8] The precise size of the blades and the angle which they present to the steam are all carefully calculated from the laws of fluid dynamics to extract the maximum proportion of the steam's energy. The changing temperature, pressure, and volume of the steam as it passes through a set of blades and from one set to the next, require changes in blade design from one end of the turbine to the other. Because the blades are all attached to the same shaft in any given turbine, they must all rotate at the same speed. Since the volume of the steam increases as its pressure declines, the size and spacing of the blades must be increased in the lower-pressure portions of the turbine

---

[8] This discussion abstracts somewhat from the complexity of modern turbines, some of which have stationary as well as rotary blading. Nevertheless, the basic principle holds.

in order that this larger volume of steam may pass through the turbine without an increase in speed.[9]

There is no theoretical reason why the efficiency of this expansion process could not be 100 per cent. There are, however, a number of practical reasons why it falls short of this figure. One important reason is friction. The movement of the steam against the turbine blades is not perfectly smooth and neither is the passage of the steam through the various nozzles through which it is carried and directed against the blades. In addition, there is usually some turbulence in the steam and these deviations from smooth, isentropic[10] flow tend to waste part of the energy in the steam. Since the amount of the frictional loss is related to the volume of steam, the losses are greater in proportion to the energy remaining in the steam in the later stages of expansion. The frictional losses are also increased by the gradually increasing amounts of moisture which occur in the steam in the process of expansion. This loss is particularly great for the lower temperature cycles since the final exhaust of such cycles contains a greater proportion of moisture. This effect of low temperatures tends to reduce the benefits of using saturated steam that are indicated by the Rankine cycle.

The effect of these various kinds of friction upon the heat cycle is illustrated by Figure A-11, which shows the energy content of the steam versus its entropy as it passes through the turbine. The result of fluid friction in the turbine is the retention of some energy in the steam which would otherwise have been delivered to the turbine blades. The steam at the exhaust pressure therefore contains more energy than it otherwise would, and this heat is wasted in the condenser instead of being turned into useful work in the turbine. In the diagram, ideal expansion is represented by the vertical path *AB*, whereas the actual expansion is along *AB'*, the deviation from the ideal being represented by the bending of the line. The energy delivered to the turbine blades is the vertical change from *A* to *B* or *B'* and the greater heat loss to the condenser in the actual cycle is measured by the higher ending point of the bent line. The increasing friction losses as the moisture content of the steam rises are represented by the increased curvature of the steam condition line as it passes the saturation curve; that is, as latent heat is

---

[9] Here, too, reality is somewhat more complicated than theory. In modern large turbines some of these volume-speed problems have been dealt with by passing the steam through two different turbines.

[10] Isentropic is, literally, constant entropy. This is the characteristic of a perfectly smooth expansion against the maximum possible load, with no heat losses and no frictional effects.

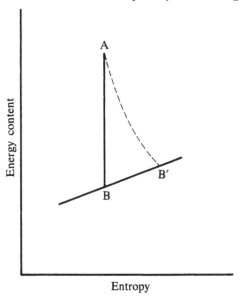

Figure A-11. Effect of non-ideal expansion in a steam engine.

lost and the steam begins to condense. This increasing loss at saturated conditions reduces the theoretical advantage of operating saturated steam cycles, since all of the expansion phase of such cycles takes place in the wet or mixture region.

Another source of loss results from leakage around the blades. The high-pressure working fluid tends to escape in all directions if unobstructed; hence, the permitted tolerances in turbine construction must be quite small. This requirement is made even more difficult because metals expand when heated. Small differences in the metallic composition of a turbine blade and its shell will cause a blade which fits perfectly into its housing at one temperature to be a misfit at another. The problem is accentuated at high temperatures and pressures because at those conditions large quantities of energy are present in small volumes of steam and the proportion of the energy which is lost through leakage may be high. A partial remedy is to increase the size of the unit for cycles which have especially large temperatures and pressures, but the problem does place a limitation on the flexibility of the engineering for steam plants.

Current best-practice steam turbines have operated for some time at about 85 per cent efficiency (reference 8, p. 74).[11] Although tending to be

---

[11] Here, efficiency is the actual work output in steam expansion divided by the ideal (isentropic) expansion work to the same pressure.

reduced by the losses attending high-pressure (small volume) steam cycles, efficiencies have been improved by the increasing use of larger-size units and by accompanying refinements in design and fabrication. Because of the increasing proportional importance of the steam leakage losses as total steam flow falls, the turbine efficiency declines at less than full load and can be increased somewhat by overload. These changes in efficiency with load are large enough to be significant in steam plant operation, so turbines are operated at full load whenever possible. Sizable reductions in power demand are usually met by taking turbines completely out of service in a large system. Also, large new units are operated at the highest practical load factor, while fluctuations in power demand are generally balanced by partial or total shut-down of older, less efficient units.

Combustion and Heat Transfer. The beginning of the electrical generation process must necessarily involve the creation of the heat energy that will eventually be changed into electricity. Rising relative prices for most fuels over the past several decades have encouraged techniques for improving combustion practices so as to permit use of low-quality fuels and for extracting the maximum proportion possible of the heat energy available in the fuel. These processes have for the most part revolved upon the purchase of the low-quality fuels by the utilities from primary suppliers and the establishment of relatively large facilities for the processing and treatment of the fuel at the plant site by the utility itself. This treatment usually involves such processes as cleaning, washing, pulverizing, etc., particularly for coal-using establishments, and increased facilities for the handling of ash and slag resulting not only from the decreasing fuel quality, but also from the strengthening of anti-air pollution legislation by municipalities.

But while the technology of combustion practice has improved substantially, there has been little or no change in basic scientific technique over the period. The principle—the addition of oxygen in the form of air to the boiler fuel at sufficient temperature and in the proper proportion for optimum chemical reaction—has remained unchanged except for the unique fuel technology involved in the nuclear fuels. The losses in energy efficiency arise from incomplete combustion or incomplete heat transfer. The combustion losses occur when part of the fuel charge is not completely burned. That is, the ash or slag, which is primarily non-combustible minerals in the fuel along with small quantities of partially burned fuel, absorbs heat from the burning fuel which it does not pass on to the steam because it is impractical to provide for this heat exchange process.

Although the slag and ash losses are the primary visible products that indicate imperfections in the combustion process they are not the only sources of combustion losses. We have already indicated that the oxygen required for combustion is usually supplied in the form of air, and since air is only about 21 per cent oxygen, there is a generous supply of other unburned substances, largely heated nitrogen, in the exhaust from the combustion chamber. In addition, more than the minimum amount of the very cheap air is usually added to the fuel in order to insure maximum combustion. It is unlikely that the economics of combustion will ever change enough to noticeably reduce these losses.

Even the attainment of perfect combustion would not ensure perfect efficiency of the boiler process, for under optimum conditions of combustion the heat produced from the fuel cannot be completely transferred to the steam. The heat transfer process must take place to a working fluid that is under high pressure and which will, during the process, reach a high temperature. It would be impossible, therefore, for the combustion gases to give up heat below this temperature, so they must leave the boiler furnace at least at this level. In addition, the fluid must be enclosed in tubes which separate it from the combustion gases and impede the transfer of heat from the gases to the fluid. Thus, the temperature of the gas must be even higher than that of the fluid if transfer is to take place at satisfactory rates. Under these conditions, there is necessarily some wastage of heat due to the fact that transfer ceases at some point before the gases give up all of the heat acquired in combustion.

In modern boiler practice, attempts at utilizing the heat that would otherwise be wasted in the stack exhaust have led to use of heat exchangers to transfer the heat to the incoming combustion air and to the feed water. These savings, instituted at the expense of the additional equipment, have been important in improving the efficiency of steam power plants. Up to a point, particularly in large, high-pressure plants, costs of these improvements are more than offset by fuel savings.

In recent practice, the boiler efficiency of the best units has run at about 90 per cent (reference 54), which is slightly above the best turbine efficiencies and far above the efficiency of the best steam power cycles. The opportunities for improving efficiency in the furnace-boiler operation are considerably lower than in the ideal cycle itself, and some of the techniques for improving cycle efficiency, e.g., the use of higher temperatures, tend to make even existing standards of boiler efficiency more difficult to maintain. At any rate there has been little improvement in the thermal efficiency of the best boilers in the post-war years, although there have been substantial improvements from an economic point of view.

Appendix B

The Supply Curve
of Thermal Efficiency

Having dealt in Appendix A with the physics of thermal efficiency, we turn here to the economics of producing thermal electric power. In particular, we are concerned with those aspects of thermal station design which introduce elements of flexibility in the consumption of fuel—and thereby, water. Since not all of the physically possible economies have already been realized, we can infer that such economies are only achieved by the use of additional amounts of other important inputs and, therefore, the costs of these substitutes weigh heavily in determining the economically desirable level of fuel economy. Our object is the construction of a supply curve of thermal efficiency. First, the purely technological aspects of the question will be presented—the development of the production function for thermal efficiency. This function, together with the economic relationships to be developed in later sections, will yield us a measure of the opportunity costs of achieving additional fuel and water economies.

THE PRODUCTION FUNCTION

Even under the most ideal circumstances, attempts to estimate the shape of a production function—particularly if it is to be made in detail—are fraught with difficulty. Some kinds of problems arise because the number of important inputs is usually very large and/or because in most industrial processes the number of outputs is also substantial. Others arise because the technical secrecy imposed by economic competition, inhibits the availability of data, particularly that portion which describes the frontier of knowledge in a field. In some segments of economic activity, the difficulties are aggravated by localized or rapidly changing techniques of production.

The situation in the electric generating industry is more favorable for estimating purposes than in most industries. Two inputs, capital and fuel, account for over 90 per cent of total annual generating costs. The level of these costs is, in turn, determined by only a handful of factors. The industry, being a regulated public utility, is considerably less secretive than most about its technology, although its status as a public utility makes it more sensitive about certain other kinds of information. Furthermore, although there are regional differences in generating technique used and, in a few companies, technical procedures which distinctively reflect the influence of a single dominant executive, the industry is remarkably standardized. Despite the relative availability of information, the complexities of investigating the technological procedures are such as to prevent a precise mathematical formulation. The industry data are not usually organized in a form useful to the economist; moreover little information is provided on the costs of equipment that is over the horizon of present use. As a result, some of the material that follows is semi-intuitive in nature. If we are to draw rational conclusions from this study, the questions raised, however, must be answered. In so doing the authors draw on the best information available to them.

The Concept of a Production Function. The technical problems mentioned above are not the only source of confusion in material dealing with the production function. All too often the definition and concepts carefully developed in texts and theory are ignored in the empirical analysis, frequently at the expense of clarity. This is a problem common to all economic research, but is sufficiently important to the outcome of our analysis to warrant a digression explaining the concept of a production function.

In standard texts on economic theory, the production function is simply defined as a functional relationship embodying *all technological knowledge* regarding the production of the commodity involved. In theoretical work where somewhat greater accuracy is useful, it can be more narrowly defined so as to exclude that technology which would use more of one input and as much or more of every other input in order to produce any given quantity of output. To illustrate, let us assume that in order to produce a given amount of electricity, we can use three

Fuel Costs

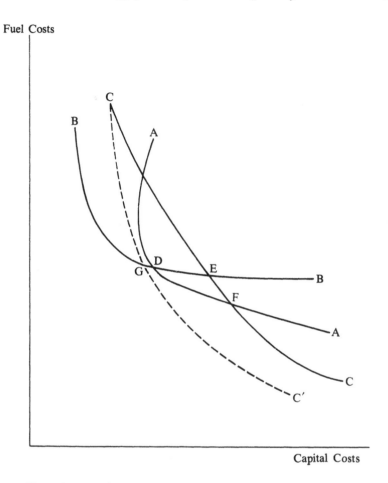

Figure B-1. Production function for electric power; all curves relate to the production of one kilowatt-hour.

Capital Costs

different techniques using different combinations of fuel and capital. One technique, say the use of the steam turbine, permits the use of various combinations of fuel and capital in the production of X units of electric energy, as indicated by the line $ADFA$ in Figure B-1. Another, say the gas turbine, permits us to use various combinations of the same factors along the line $BDEB$, while still another, say atomic power, offers the possibility CEC. Now the specific combination of factors that will, in fact, be used, depends on the prices of fuel and capital that exist at the time the decision is made (or that are expected to prevail over the period that must elapse before the decision can be changed). But regardless of the price of any of the inputs (as long as the price is not negative) no rational producer would operate at any point which lay both above and to the right of another point, since that would involve using both more fuel and more capital in producing the power.[1] Thus, while all the methods are technologically feasible, only those points on the path *BDFC* are meaningful as long as we employ the customary assumptions of rationality and efficiency.[2]

Using this approach, we can now define the production function as the functional relationship which describes *all* of the "most efficient" technology known to the producer, where efficiency is defined without recourse to any price information. Regardless of which technique is currently in use, the shift to another method of production which was known but not utilized previously does not mean that the production function has changed. It only implies that changing economic conditions have made it more profitable to produce at another point on the function. A change in the relative prices of fuel and capital might very well cause a utility to shift from conventional to atomic power, and yet such a move would not necessarily imply technological change as the term is properly used in economic theory.

---

[1] Of course, a firm which built a plant before the discovery of a new technique might have a plant which operates on segment *FA* and, having built it, might choose to operate it in preference to a new plant so long as its operating costs were less than total costs of a completely new one. That, however, does not blunt the force of the argument.

In most empirical studies of production functions it is precisely this "short-run" relationship which is evaluated. That is, the nature of actual production input data limits the researcher to observations of variations of factor input and production *given* the present stock of capital. A change in relative factor prices will cause an adjustment of inputs but only within the limitations of the current plant. Furthermore, if the researcher tries to extend the period of observation to permit a "longer-run" adjustment, he also runs the risk of giving the function an opportunity to shift— that is, he permits the possibility of true technological change. The issue is further complicated by the dynamics of the situation, i.e., by the fact that at any point in time there will be many production units at any number of points in the "short-run" adjustment process. All of these objections can be avoided by use of engineering data, if those data are available, though this then requires the introduction of a procedure for relating this engineering data to the economics of the question.

[2] It should be noted that we have intentionally avoided for the moment all of the dynamic problems that may arise, e.g., the possibility that while the movement from *B* to *D* may be made instantaneously, movement further "southeast" along the curve may require large-scale process readjustment, which has secular implications.

Expressed symbolically, the argument is as follows: Define a production function

$$P = g(F, C) \qquad \text{(B-1)}$$

where P = Production
F = Fuel
C = Capital.

Given factor prices P_f and P_c and product price P_e, the firm tries to maximize the expression

$$\Pi = p_e P - p_f F - p_c C \qquad \text{(B-2)}$$

subject to:

$$P = g(F, C). \qquad \text{(B-3)}$$

Solving, we get the customary expressions

$$\frac{\dfrac{\partial g}{\partial F}}{P_f} = \frac{\dfrac{\partial g}{\partial C}}{P_c}. \qquad \text{(B-4)}$$

As P_f and P_c change relative to each other, the appropriate levels of $\partial g/\partial F$ and $\partial g/\partial C$ must also change to maintain the equality and, thus, the particular point on the production function at which production takes place must also change. Note that this is true regardless of whether or not the *level* of production changes. Conversely, this change in the point at which production takes place does not imply a change in the form of the production function. As long as all the possibilities at our disposal, *all represented in the function* $P = g(F, C)$, remain the same, the production *function* does not change, even though the particular *technique* does.

This does not mean that all process changes are due solely to economic factors. One process might benefit from an invention which does not apply to another, and thus change the shape of the function. For example, referring back to Figure B-1, if a technological change were to shift CEC to CGC', the new production function would be BGC' and as a result process $ADFA$ would no longer be used. This kind of shift of process represents a true technological change, but the distinctive feature of the technological change was not the shift in process but the change in the known techniques for producing the unit of output. Changing techniques over time are not sufficient evidence to prove the case for technological change in the sense of shifts in the production function, although either type of change can serve to reduce the cost of producing any given output.[3]

The distinction between the movements along, and shifts of, the production function are not idle ones, since they have vastly different implications for any project concerned with projecting the future path of technology. In the one case the determining elements of the study must be economic—the evaluation of the effect of changing input prices upon the method of production—and in the other it must be technologic—the likelihood of certain improvements in technique. As in most cases in the real world, the electric industry's growth has been typified by both types of changes, but the ability to distinguish between these two elements should improve our ability to analyze the patterns of change.

To the extent that changing techniques of production merely reflect changing economic incentives, the necessity of formally predicting the rate of invention and innovation is eliminated. In such circumstances, the projected movements of the principal economic factors will be the principal determinants of economic change. If, on the other hand, the rate of change of technological knowledge is expected to be rapid, it may be possible to neglect the effect of changes in the economic parameters without introducing any important error. It is obvious that the real world is not as polar as all this, and that we are likely to find any number of possible intermediate positions more tenable than either of these extremes. Under these circumstances, the problem reduces to measuring the kinds of technical change that might be expected both from moving along the curve, on the one hand, and shifting the production function on the other. While it is very difficult at our present state of knowledge to predict the course of technological change, we can get an idea of the probable importance of innovation by investigating the nature of the most likely innovations and their effects on production techniques. Similarly, we can see what the effect of certain changes in the relative prices of factor and product inputs would be. The smaller the effect of major changes in factor prices on production techniques, the greater will be the importance of innovations, and conversely.

[3] It should be noted that, while either type of change can serve to reduce costs, technological change will always reduce costs, given factor prices, while changes in factor prices will have either effect, depending on the shape of the curve and the nature of the price change. Similarly, changing prices does not by itself increase real gross national product, while changing technology, barring unusual factor supply conditions, always will.

One major obstacle in work on the production function is the practical difficulty of distinguishing between movements along the function and shifts of the function itself. Because such an investigation usually requires extensive knowledge of industry, students of technological change have tended to ignore the difference between these two types of movement. In discussing entrepreneurial responses to changing conditions, the distinction between a change induced by economic phenomena and one induced by a change in the state of the arts may not be an important one. Nevertheless, it is important in a projection of the future not to overestimate the amount of change that takes place through changes in knowledge. For example, we have noted that historically the level of thermal efficiency in steam generating plants has risen substantially through a steady increase in temperatures and pressures. Yet we have also pointed out that the possibility of such improvements in efficiency could readily be predicted from the theory of steam cycles available to engineers fifty years ago. We cannot be sure that technological change has taken place *merely* because efficiency has improved, because an increase in the relative cost of fuel as compared with the cost of capital will lead to efficiency increase also.[4] Even the much-ballyhooed move to supercritical steam cycles, using metals never before used in steam generating equipment, does not imply a change in knowledge in any but the narrowest sense, since both the theory of the steam cycle and fabrication experience with the new metals in other fields was well advanced beforehand.

THE PLAN FOR DEVELOPING THE
SUPPLY CURVE OF THERMAL EFFICIENCY

In Appendix A we developed a set of curves in Figure A-7 which related thermal efficiency to temperature and pressure in a simple, ideal, Rankine cycle.[5] If temperature and pressure are considered to be factors of production, these relationships are, in effect, a production function for thermal efficiency in terms of those factors. Symbolically, the form of such a production function is:

$$H = h(T, P) \qquad \text{(B-5)}$$

[4] On the other hand, we are not arguing the reverse, either. There have actually been substantial technological changes in the industry that helped to make these improvements in efficiency economically possible. But the improvements do not follow automatically from the fact of changing techniques.

[5] "Simple" refers to the fact that there was no reheat or feedwater heating in the cycle; "ideal" refers to the neglect of mechanical inefficiencies.

where H is thermal efficiency, T is temperature and P is pressure. The curves are a two-dimensional representation of this relationship exactly analogous to the more conventional production isoquants of economic theory. For each level of H, say H_i, we have a two-dimensional curve, $h(T, P) = H_i$. The bounding line on the left of the curves represents the limitation imposed on our production function: that it include no portion of the production possibilities that entailed the use of more of one input and no less of another.[6]

In this appendix the simple relationships of Appendix A will be modified by the introduction of the effects of reheat and regeneration. Since the number of possible isoquants rises very sharply with each possible modification in initial design,[7] clarity demands that some of these modifications be eliminated on the grounds that they are unlikely, i.e., economically unreasonable.

The development of these isoquants does not, however, definitely solve the problem of the production function. The factors of production arising out of this formulation are necessarily quite abstract. There is certainly no market for the purchase of "temperature" or "pressure" and if these concepts are to have any economic meaning they must be translated into marketable factors. Hence, this section of this appendix will be devoted to the development of a transformation function between capital and the "primary" factors, temperature and pressure. More specifically, this implies the construction of a schedule indicating the alternative "quantities" of temperature and pressure which can be "bought" with any given amount of capital. Symbolically, we can represent this function in the following manner:

$$C = c(T, P). \tag{B-6}$$

Like the curves representing the production function, this function can be represented two-dimensionally. In this case, however, the curves take the form of transformation curves, which depict the manner in which temperature and pressure can vary, given any level of capital expenditures.

Unlike the efficiency isoquants, these transformation curves are concave to the origin, reflecting the increasing relative difficulty of

[6] The reader with experience in visualizing such curves may be disturbed by the fact that they do not appear to have the conventional shape; i.e., convex to the origin. This deviation from normality arises from the use of semilogarithmic scales, which was required by the relative magnitude of the variation in the two factor inputs.

[7] For example, if there are p possible reheats and q possible stages of feedwater heating, there will be pq possible isoquants for each level of thermal efficiency.

raising temperatures at the expense of reducing pressures and vice versa. The transformation curves, like the iso-efficiency curves, represent only the "efficient" possibilities. This implies that they represent only the *maximum* amounts of temperature and pressure that can be "bought" with the given amount of capital. Again, the efficiency condition is that no point on the curve may offer the possibility of less of one "factor" without simultaneously offering the possibility of more of the other.

At this point in our presentation we have developed two relationships:

$$H = h(T, P) \qquad \text{(B-7)}$$
$$C = c(T, P).$$

Our objective in this procedure is the construction of a supply schedule of the form

$$H = s(C), \qquad \text{(B-8)}$$

representing the costs of increasing efficiency to set against the demand curve

$$H = d(F), \qquad \text{(B-9)}$$

which represents the potential fuel savings from changes in thermal efficiency. In order to eliminate T and P in equation B-7, we need one more independent relationship. That relationship follows readily from the maximization conditions: the requirement that each dollar of capital "produce" the maximum amount of thermal efficiency obtainable. It can be readily demonstrated that this condition is equivalent to the relation:

$$\left(\frac{dP}{dT}\right)_H = \left(\frac{dP}{dT}\right)_C. \qquad \text{(B-10)}$$

Eliminating P and T between the three equations yields us the desired supply curve.

The same solution can be shown graphically. In Figure B-2 are shown iso-efficiency curves depicting the different ways in which temperature and pressure can be combined to produce various levels of efficiency. On the same graph are the curves indicating the various ways in which a given total amount of real capital resources can be used to produce temperature and pressure. Given these curves and the maximization motive, we can see that while (say) the curve representing $135/kw of capital resources crosses both the 38.30 per cent and 38.75 per cent efficiency lines, the maximum efficiency producible with those resources is achieved by the combination of temperature and pressure to produce 39.20 per cent efficiency. Any move in either direction off the point of tangency will reduce the efficiency attainable. The dotted line, which connects all the points of tangency between the two sets of curves (including those curves reflecting fractional efficiencies not depicted for ease of presentation) is in effect the desired supply curve for thermal efficiency, since it is the locus of all points of maximum efficiency attainable with the given amounts of resources. In conventional form, this curve is depicted in Figure B-3.

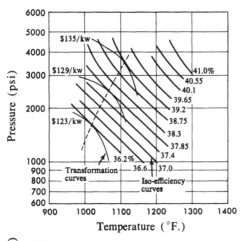

Figure B-2. Iso-efficiency curves and transformation curves, showing points of maximum efficiency.

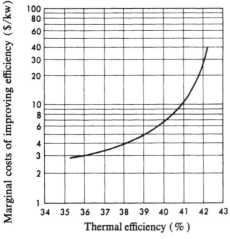

Figure B-3. Supply curve of thermal efficiency.

With this background, we can now turn to an investigation of the nature of the supply of thermally efficient power stations. The question before us is to determine the costs of marginal improvements in fuel utilization, but even before we turn to an empirical determination of the costs involved, we can cast some conceptual light on the nature of the results we will find by drawing upon some of the more abstract physical data we developed in Appendix A.

We indicated in Equation B-10 that the slope of the production isoquants could be written as:

$$\left(\frac{dP}{dT}\right)_{H_i}.$$

That is, the slope of the curve is the rate of change of pressure with respect to a temperature change, holding the level of thermal efficiency constant. Stated another way, it is the rate at which pressure increases can be substituted for temperature decreases without changing the level of thermal efficiency.

The question of primary interest to us, however, is not movements along the constant efficiency curves, but rather movements between curves. Here, too, the patterns can be determined visually, but since they are somewhat more complex we will first lay the logical groundwork for them. It has been indicated that the functional relationship linking efficiency, temperature, and pressure is

$$H = h(T, P). \tag{B-5}$$

Knowing this, we can deduce that

$$dH = \frac{\partial H}{\partial T} dT + \frac{\partial H}{\partial P} dP. \tag{B-11}$$

If we stay on any given isoquant, $dH = 0$, so that

$$\frac{\partial H}{\partial T} dT = -\frac{\partial H}{\partial P} dP \tag{B-12}$$

then:

$$\frac{dP}{dT} = -\frac{\partial H/\partial T}{\partial H/\partial P}. \tag{B-13}$$

Since efficiency increases with increases in pressure and temperature,

$$\frac{\partial H}{\partial P} > 0 \quad \text{and} \quad \frac{\partial H}{\partial T} > 0,$$

so that dP/dT must be less than zero.

In effect, then, dP/dT is the ratio of the marginal physical productivities of the two factors. If dP/dT rises (in the negative direction, i.e., becomes a larger negative number), it implies that the marginal productivity of P is declining relative to that of T. If dP/dT declines, the reverse is true. If thermal efficiency were subject to constant returns to scale, dP/dT would be constant as long as factor proportions remained the same no matter what the level of thermal efficiency. If prices were constant, average and marginal costs of thermal efficiency would remain constant as well. These relationships are explained later.

It is clear, however, that the production of efficiency is *not* a constant returns effort. We have already indicated that the maximum theoretical efficiency is substantially less than 100 per cent whereas there are no known upper limits to temperature and pressure. Doubling P and T from 200 psi and 400°F. to 400 psi and 800°F. increases the theoretical Rankine cycle efficiency from 33 per cent to only 38 per cent. Even doubling the absolute temperature from 860°R. to 1720°R. (400°F. to 1260°F.) at these same pressures increases the theoretical efficiency by only slightly over one fourth, to 42 per cent. Furthermore, the closer the approach to high efficiency, the smaller will be the increase in fuel saving associated with these factor increases.

Not only do proportional increases in P and T not produce proportional increases in thermal efficiency, but there has also been a continuing shift in the relative "productivity" of changes in P and T in advanced cycles. To see this, look at Figure B-4. The lines AB, $A'B'$, $A''B''$, etc., reflect constant rates of exchange between P and T; i.e., they imply constant relative "prices" or costs of unit changes. Each line represents the different amounts of P and T that can be bought with a fixed sum of money. The slope of the line represents the ratio of the "price" of T to that of P. The most (economically) efficient combinations of T and P required to produce higher efficiencies given the relative price ratio AB are given by the points of tangency K, L, M, and N. Notice that when moving from 36.2 per cent to 37.4 per cent efficiency (point K to point L), we would raise T from 1,005 to 1,040 and P from 1,700 to 2,325. To make the further improvement to 38.3 per cent efficiency, however (to point M), requires a greater absolute increase in temperature per unit increase in pressure. This tendency increases still further as efficiency improves to 39.2 per cent. To see this another way, we have drawn the line OC through tangency points K and L. Thus, we can see that points M and N draw farther and farther away from the line OC.

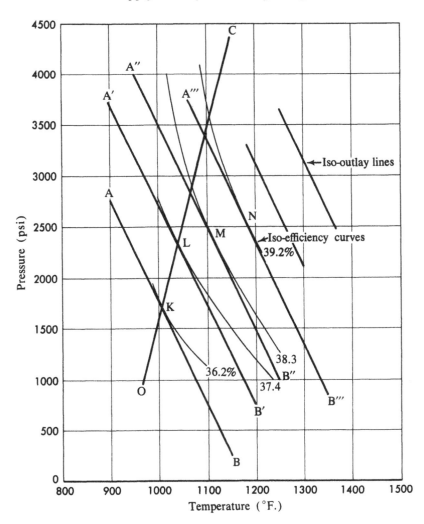

Figure B-4. Iso-outlay lines and iso-efficiency curves showing decline of dP/dT as P and T rise proportionately. (Single reheat to initial temperature.)

All this, however, assumes that price ratios remain the same. If the price ratio also shifts in favor of temperature (i.e., the price of temperature increases falls relative to the cost of pressure increases), this process of substitution would be encouraged. Temperature increases would not only be more productive than pressure increases in increasing thermal

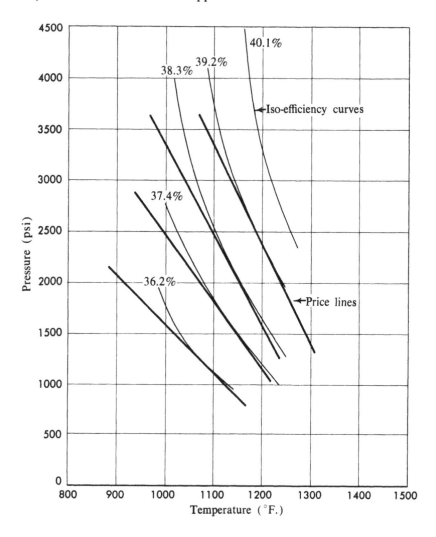

Figure B-5. Iso-outlay lines and iso-efficiency curves showing relative cost decline of *P* increase more than compensating for lesser effectiveness in increasing efficiency. (Single reheat to initial temperature.)

efficiency, they would also be relatively cheaper. On the other hand, if temperature increases proved to be more costly, the process of substitution would be slowed and possibly even reversed. Under these circumstances, the relative cheapness of pressure increases would more than compensate for their lesser effectiveness in increasing efficiency—a situation which has, in fact taken place historically and is illustrated in Figure B-5.

So far we have dealt only with relative factor *price* ratios. However, it is also possible to discuss inputs with explicit reference to price by developing a "transformation curve." This curve represents the various combinations of factors that will require the expenditure of a given amount of money. Like the slope of the iso-efficiency curves, the slope of the transformation curve can be written as the ratio of two other marginal quantities—in this case, the ratio of the marginal costs of increasing temperature and pressure.

$$\left(\frac{dP}{dT}\right)_C = \frac{-\partial C/\partial T}{\partial C/\partial P}. \tag{B-14}$$

Along any given transformation curve—for any given quantity of capital expenditure—the slope becomes steeper (more negative) the higher the temperature produced. The marginal costs of increasing temperatures rise and the marginal costs of lower pressures fall, increasing the magnitude of dP/dT. Moving in the other direction, to lower temperatures while increasing pressures decreases the slope.

All this pertains only to movements along a given transformation curve. In discussing the supply curve of thermal efficiency, however, we are primarily interested in discerning the potential increases in efficiency which would result from a larger expenditure of resources. From that point of view our primary concern is the differences between the various transformation curves, which measure that expenditure of resources. If the costs of increasing pressure and temperature by one unit were to remain constant over the entire range, successive transformation curves would all be equidistant. With this condition existing in the transformation curves, and constant returns in the iso-efficiency curves, we would find that efficiency could be produced at constant average cost.

Unfortunately, these conditions do not hold in either case. Efficiency is not subject to constant returns, and both temperature and pressure increases are subject to increasing cost. Because of the technical characteristics of the metals involved, this cost not only increases steadily at high temperatures, but the rate of increase in costs increases so that marginal temperature increases in the 1,200°F. range are substantially more costly than increases in the 1,050°F. range. In the case of pressure increases, the rising costs are much more modest—only slightly greater than would be indicated by constant costs.

The net result of these cost increases is twofold. If the factor prices had increased, but had only increased at the same rate, we would find that successive transformation curves had the same shape but were successively closer and closer together. This sort of situation is shown in Figure B-6. But as has been indicated, these cost increases are not proportional, so we will find that the successive transformation curves change their shape as well as their distance from preceding curves. Since the marginal costs of increasing temperature are greater than those for increasing pressure, we will find, in particular, that successive increases in capital outlays will buy less of an increase in temperature than they will of an increase in pressure. This implies that the differences between successive transformation, or iso-outlay curves, will become smaller more rapidly in the direction of the temperature axis than they will in the direction of the pressure axis. Stated another way, along a line representing proportional increases in factor inputs (constant proportions) we will find that the slope of the iso-outlay curves will become steadily steeper (more negative). This can be seen in Figure B-7.

Let us now look at the implications of these findings for the shape of

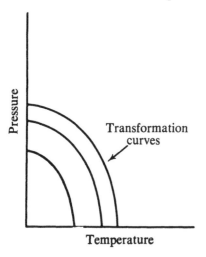

Figure B-6. Transformation curves (schematic) showing proportional increase in cost of increasing temperature and pressure.

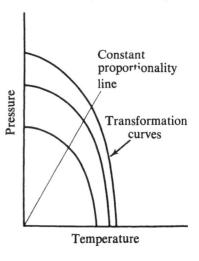

Figure B-7. Transformation curves showing more rapid rise in cost of increasing temperature than of pressure.

the supply curve of thermal efficiency. If efficiency could have been produced at constant returns but temperature and pressure costs increased proportionately, the slopes of the two sets of curves along any ray out of the origin would have remained the same. Under those conditions, successive increases in thermal efficiency would have become successively more expensive, but the ratios between temperature and pressure would have remained the same. Previously, we found that the marginal rate of substitution between temperature and pressure favored shifting to relatively greater inputs of temperature (given constant price ratios) in order to avoid the retarding effects of lower marginal productivity of pressure. We now find, however, that the price ratios tend to block this avenue of escape from declining productivity, thus accentuating the increasing cost of improving thermal efficiency.

All of these factors combine to increase the slope of the supply curve of thermal efficiency as efficiency increases. (See Figure B-8.) The solid iso-efficiency curves, 40-40 and 41-41, represent the shape the curves would have if constant returns prevailed. A ray through the origin, *OZ* cuts the two curves at points of equal slope, corresponding to the budget line (which in this example is based on constant price ratios) tangents, *A A* and *BB*. Under these conditions, an increase in outlays would keep factor proportions constant. If, however, the marginal productivity of *P* declined more rapidly than would be expected from constant returns considerations, the curve for 41 per cent efficiency might look like the dashed curve, 41-41', instead of the solid one. If the temperature-pressure price ratio remained the same, the optimal point of operation would now be *M*, which involves a relatively greater use of *T*, and a lesser use of *P* than was indicated by the former equilibria. In addition, however, the 41 per cent curve is now tangent to the *CC* budget line instead of the *BB* budget line, indicating that the same level of thermal efficiency can only be obtained by a greater outlay of funds. In other words, the result is an increased cost of efficiency due to diminishing returns.

Now let us imagine that instead of remaining the same as temperatures and pressure increase, the price of temperature increases while the price of pressure does not change. Looking at the budget line, we can see that the same amount of resources will now only permit use along the *BB'* line. Using that price ratio, we now need a much larger amount of resources, along the line *DD*, in order to produce at 41 per cent efficiency (at *N*). The net result of both the decreased productivity of pressure and the higher price for temperature is to increase the cost of producing

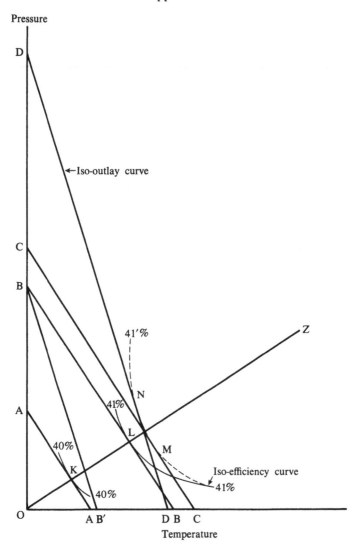

Figure B-8. Iso-outlay and iso-efficiency curves (schematic) showing returns to scale.

the given level of efficiency. The decreased productivity of pressure shifts factor proportions in favor of T; the increased price of T shifts proportions in favor of P. In this case the net effect is to increase the use of T. If the increase in price had taken place without the change in

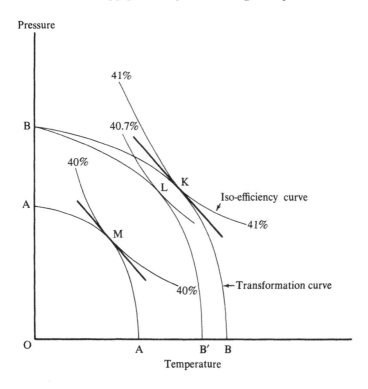

Pressure

Figure B-9. Transformation curves and iso-efficiency curves
(schematic) showing change in price ratios.

the shape of the curve, the shift in proportions would have been even
greater.

So far we have taken the price change as exogenous—as dictated to
us by outside considerations. The underlying reasons for the price
change can be seen in Figure B-9, which shows the way in which movements in the transformation curve shift the point of equilibrium.
Increasing costs of producing temperature are reflected in the transformation curve by a smaller intercept on the T axis, as from B to B'
(to A), increasing the absolute magnitude of the slope of the curve and
shifting its point of tangency, K to L (to M), toward a greater use of the
cheaper resource. If the change in the transformation curve takes place
through a rise in the cost of temperature increases while the cost of
pressure increases remains constant, the changing slope has the double
effect of changing factor proportions and simultaneously shifting the

point of tangency to a lower iso-efficiency curve. The shift from *BB* to *BB'* reflects such an increase in the cost of temperature, shifting the point of tangency from *K* to *L*.

Thus we can see that the properties of the underlying production and transformation curves all point to a sharply rising supply curve for thermal efficiency. Short of a major technological change in the future, these forces toward higher costs will become increasingly strong as efficiency increases.

Appendix C

References and
Literature Citations

1. Andrew, J. D., P. H. Kock, and E. A. Pirsh. "Operating Experience with the First Commercial Supercritical-Pressure Steam Generator at Philo," *Proceedings of the American Power Conference*, Vol. 20, 1958.
2. Archbold, M. J., F. V. Miholits, A. Leidner, and C. E. Person. "Economic Sizing of Condensers through the Use of the Digital Computer," *Proceedings of the American Power Conference*, Vol. 22, 1960.
3. Babcock and Wilcox Company, The. *Steam: Its Generation and Use.* New York, 1955.
4. Baker, D. R. "Selecting Cooling Towers for Condensing Steam Turbines," presented at 1961 Annual Meeting of the American Society of Mechanical Engineers.
5. Baker, R. A. "Mercer Generating Plant—The Case for 1,100 F. Steam," *Proceedings of the American Power Conference*, Vol. 24, 1962.
6. Barnes, H. C., T. T. Frankenberg, and C. P. Lugrin. "Hyperbolic Cooling Tower Highlights the Feature of AEP System's Big Sandy Plant," *Power Engineering*, December 1962.
7. Bartels, J. "Supercritical Pressure Steam Power Cycles," *Proceedings of the American Power Conference*, Vol. 16, 1954.
8. Bartlett, R. L. *Steam Turbine Performance and Economics.* New York: McGraw-Hill Book Co., Inc., 1958.
9. Berg, B., R. W. Lane, and T. E. Larson. "Water Use and Related Costs with Cooling Towers," *Journal of the American Water Works Association*, March 1964. Also published as Circular 86, Illinois State Water Survey (1963).
10. Bethlehem Steel Company. *Quick Facts about Alloy Steels*, undated.
11. Birget, C. D., and J. H. Kline. "Which Enclosure Shall It Be?" *Proceedings of the American Power Conference*, Vol. 21, 1959.
12. Brooke, M. "Tables Speed Cooling Tower Calculations," *The Refining Engineer*, November 1947.

13. Campbell, C. B., C. C. Franck, Sr., and J. C. Spahr. *The Eddystone Super-pressure Unit,* Westinghouse Electric Corporation, undated.
14. Carlson, Sune. *Study on the Pure Theory of Production.* New York: Augustus M. Kelley, Reprints of Economic Classics, 1956.
15. Carr, L. H., and W. Crawford. "The Case for a 4500 psi Industry Standard," *Proceedings of the American Power Conference,* Vol. 20, 1958.
16. Curran, R. M., and A. W. Rankin. "Application of High Temperature Metals to Modern Large Steam Turbines," *Proceedings of the American Power Conference,* Vol. 18, 1956.
17. Davies, R. W., and G. C. Creel. "Economics of the Selection of 2500-psig Double Reheat for a 300-Mw Unit," *Proceedings of ;the American Power Conference,* Vol. 25, 1963.
18. Edison Electric Institute, *Statistical Yearbook,* 1951-64.
19. *Electrical World,* July 20, 1959.
20. Elston, C. W., and J. E. Downs. "The Future of the Steam Turbine Cycle," *Proceedings of the American Power Conference,* Vol. 19, 1957.
21. Fairchild, F. P. "Eight Years of Experience with Austenitic-Steel Piping Materials at Elevated Steam Conditions," Paper No. 56-A-181, presented at 1956 Annual Meeting of the American Society of Mechanical Engineers.
22. Federal Power Commission, *Electric Power Requirements and Supply of the United States, by Regions, Present and Future to 1980.* Washington: April 1958.
23. ———. *Steam-Engine Plant Construction Cost and Annual Production Expenses.* Washington: Government Printing Office, 1951-62.
24. ———. *Water Requirements of Utility Steam Electric Generating Plants.* Washington: Federal Power Commission, 1954.
25. Fleischman, J. J., A. H. Gibeling, A. Mergy, and E. R. Miller. "A Comparative Study of a Large Steam Turbine Application for Supercritical and Conventional Pressures," *Proceedings of the American Power Conference,* Vol. 18, 1956.
26. Foster Wheeler Corporation. *Cooling Towers Bulletin,* CT-57-1, 1957.
27. Franck, C. C., Sr. "Steam Turbine Developments," *Proceedings of the American Power Conference,* Vol. 16, 1954.
28. ———. "Superpressure Steam Turbines," *Proceedings of the American Power Conference,* Vol. 19, 1957.
29. Frankenberg, T. T., and J. A. Tillinghast. "Symposium on the Commercial Operation of the Breed Plant: Thermal Cycle Equipment and Performance," *Proceedings of the American Power Conference,* Vol. 23, 1961.
30. General Electric Co. Personal communication.
31. Gill, N. F., and N. D. Flack, "Symposium on Avon Unit No. 8—A Supercritical Plant: Initial Operation," *Proceedings of the American Power Conference,* Vol. 23, 1961.
32. Goitein, E. E. "Selection and Application of Cooling Towers in Steam-Electric Stations," *Heat Engineering,* Vol. 32, Nov.-Dec. 1957.

33. Hamilton, T. H., Fluor Products, Inc. Personal communication.
34. Hansen, E. P., A. C. Holmes, and J. J. Fleischman. "High Temperature Steam Turbine Design Improvements from Operating Experience," *Proceedings of the American Power Conference*, Vol. 20, 1958.
35. Harlow, J. H. "Engineering the Eddystone Plant . . . for 5000 lb. 1200-Deg. Steam," Paper No. 56-A-165, presented at 1956 Annual Meeting of the American Society of Mechanical Engineers.
36. Jakob, M., and G. A. Hawkins. *Elements of Heat Transfer*. 3rd ed. New York: John Wiley and Sons, Inc., 1957, pp. 173ff.
37. Keenan, J. H., and F. G. Keyes. *Thermodynamic Properties of Steam*. New York: John Wiley and Sons, Inc., 1936.
38. Kelley, A. G., and N. R. Lawless. "The Economic Sizing of Cooling Towers," *Proceedings of the American Power Conference*, Vol. 24, 1962.
39. Larson, T. E., Illinois State Water Survey, Personal communication.
40. Lichtenstein, J. "Performance and Selection of Mechanical Draft Cooling Towers," presented at the 1943 Spring Meeting of the American Society of Mechanical Engineers.
41. Lockhart, F. J., J. M. Whitesell, and A. C. Catland, Jr. "Cooling Towers for the Power Industry," *Proceedings of the American Power Conference*, 1955.
42. Marley Co., The. *Marley Engineering Manual for Industrial Cooling Towers*.
43. ———. Personal communication.
44. McKelvey, K. K., and M. Brooke. *The Industrial Cooling Tower*. Amsterdam: Elsevier Publishing Co., 1959.
45. Menson, J. L., W. L. Harding, and E. P. Petit. "Predicting Tube Life in High Temperature Boiler Installations," *Proceedings of the American Power Conference*, Vol. 22, 1960.
46. Miller, E. H., and B. M. Cain. "Steam Turbine-Generators of the 1960's," *Proceedings of the American Power Conference*, Vol. 21, 1959.
47. Miller, E. H., and A. Sidun. "Economic Determination of Condenser and Turbine-Exhaust Sizes," Paper No. 53-A-94, presented at the 1953 Annual Meeting of the American Society of Mechanical Engineers.
48. "Modern Plant Survey," *Power*, published annually, 1953-1958.
49. National Coal Association. *Steam-Electric Plant Factors, 1956-1961*. Washington, 1962.
50. Patterson, M. M., E. V. Pollard, and W. B. Wilson. "Economics of Higher Pressures and Temperatures for Steam Turbines in Industrial Plants," *Proceedings of the American Power Conference*, Vol. 19, 1957.
51. Petersen, H. J. "The Economics of 2400 psig versus 3500 psig for Large Capacity Units," *Proceedings of the American Power Conference*, Vol. 25, 1963.
52. Pfeiffer, E. L. "Preliminary Cooling Tower Selection," *Heat Engineering*, March-April, 1949.

53. ———. "What Size Cooling Tower?" *Chemical Engineering*, Vol. 56 (4), 1949.
54. Powell, E. M. "Engineering the Eddystone Steam Generator for 5000 psig, 1200 F Steam," presented at the 1956 Annual Meeting of the American Society of Mechanical Engineers.
55. Ritchings, F. A., and S. Crocker. "Central Station Piping Trends," *Proceedings of the American Power Conference*, Vol. 18, 1956.
56. Ritchings, F. A., and A. W. Lotz. "Economics of Closed versus Open Cooling Water Cycles," *Proceedings of the American Power Conference*, Vol. 25, 1963.
57. Samuelson, Paul. *Economics: An Introductory Analysis*. New York: McGraw-Hill Book Co., Inc., 1948.
58. Seelye, H. P., and W. W. Brown. "The Economy of Large Generating Units," *Proceedings of the American Power Conference*, Vol. 16, 1954.
59. Sheppard, R. "Operating Experience with the First Commercial Supercritical-Pressure Steam Turbine Built for the Philo Plant," *Proceedings of the American Power Conference*, Vol. 20, 1958.
60. Smith, A. F., and R. E. Bovier. "Hyperbolic Cooling Towers with Reservoir Storage of Makeup to Serve the Proposed Keystone Generating Station," *Proceedings of the American Power Conference*, Vol. 25, 1963.
61. Teitelbaum, Perry D. *Nuclear Energy and the U.S. Fuel Economy 1955-1980*. Washington: National Planning Association, 1958.
62. Ulmer, M. S. *Capital in Transportation, Communications, and Public Utilities: Its Formation and Financing*. Princeton: Princeton University Press, 1960.
63. Weir, G. E., and J. F. Brittain. "Economic Features in the Selection of Circulating Water Supplies for Electric Generating Stations," *Proceedings of the American Power Conference*, Vol. 24, 1962.
64. Wilson, R. M., and R. and F. B. Foley. "Metallurgical Requirements of Metals for Steam Service Above 1000° F.," *Proceedings of the American Power Conference*, Vol. 18, 1956.